The World in Bullshit:
Fooling most of the People, most of the Time

Will Rowlands

First Edition 2022

Copyright under exclusive licence from Will Rowlands.

The rights of Will Rowlands have been asserted in accordance with the Copyright, Designs and Patents Act 1998.

No part of this book may be reprinted or reproduced or utilised in any form or by any electronic, mechanical, or other means, now known or hereafter invented, including photocopying and recording, or in any information storage or retrieval system, without permission in writing from Will Rowlands.

ISBN-13: 978-1-913822-20-0

Acknowledgements

So many people lost their minds during 2020 and while it was never personally an issue keeping sane in the face of such transparent bullshit, a bit of moral support was always helpful. Thanks to all the people who offered such support, however minor you think it might have been. God knows there's been enough unpleasantness and every little gesture helps morale.

It would be impossible to pierce the fog of lies without a little helping hand.

The golden rule is not to trust a particular brand or source or individual but to believe in reasons and arguments based on facts. That said, one man stands out for consistently delivering concise information that cuts through the lies and exposes the agendas in a way that is simple to understand. Thank you to Mark Windows (www.windowsontheworld.net) for your efforts to explain the global action plan and for doing it in a humorous way. I haven't missed one of your Sunday shows in years.

Again, a back handed thank you needs to go to the people trying to drive this global change. You've scored an own goal with your rapacious plans by awakening millions of good people, many of whom I can now call my friends. Your plans will fail and your ambitions have only served to show how many good people there are compared to the sick, elitist 0.0001%.

Many thanks to Ken and Reinette for giving me the opportunity to publish this book and, of course, to Joan for correcting all my errors.

This book is dedicated to the memory of Niki Raapana,

You saw the whole of the moon.

PART ONE

1.1

The Meteoric Rise of Greta Thunberg

"We accept the reality of the world with which we're presented. It's as simple as that."
Christof (from The Truman Show)

August 2018

A fifteen-year-old Swedish girl sits alone outside the Swedish Parliament building with a sad face and a banner with the words 'Strike For Climate' written on it.

A picture of her protest is posted to social media and the very next day other children join in. The picture is widely shared and quickly goes viral across the world as the schoolgirl gains thousands of followers on social media. Within a week, her protest attracts the attention of local news outlets and then international news outlets. Public appearances follow where people flock to catch a glimpse of the new Green messiah. Before long, school kids are campaigning for the environment in hundreds of cities across the whole world, skipping school en masse to attend protests, something that is strangely encouraged by the media and even some teachers. Reporters hang on her every word despite her tender age and her slightly wooden script reading – probably because she suffers from Aspergers – but she shrugs this aside in her energetic efforts to further the cause, the noblest cause of all and one that cannot be questioned.

Despite her complete lack of scientific knowledge from not attending school, she is certain of one thing; carbon dioxide is evil and must be reduced or the world will suffer terrible consequences. She wants us to act fast because the situation is urgent. We must act as if the house is on fire and understand that this is the sixth mass extinction event! She does not travel by plane or eat meat, use plastic straws and a lot more besides. World leaders adore her and are keen to be photographed with her. Her face appears on the cover of *TIME* magazine. One day, she travels across the Atlantic on a carbon-neutral yacht, reaching the UN headquarters where she delivers a stern lecture to the assembled Presidents, Prime Ministers and other dignitaries of the world about the urgent need to change their ways.

And that is how Swedish polymath Little-Miss-How-Dare-You rose to fame. Seeing as all this occurred organically and through grassroots activism, it just goes to show anybody from any background can overcome life's obstacles and capture the whole world's attention if the cause is worthy enough. It is a wonderful story and one that can inspire us all.

Well, this is what we are supposed to believe. Millions did. Please excuse the slightly sarcastic tone but you probably worked this out by yourself anyway. This is a book about bullshit after all.

Returning to the present...

Of course, we are talking about the meteoric rise of Greta Thunberg, 'Meteoric' is probably the wrong word to use in this context and not just for the fact none of us have ever seen a meteor rise. Even if meteors did rise, it would still be the wrong word. There is a far more truthful description, so let's call it what it is, Greta Thunberg's stage-managed rise to fame.

So what happened between the summer of 2018 and Greta's 'How Dare You' tirade at the United Nations (UN)?

It turns out that Greta had a few helping hands to nudge her in the direction of super-stardom. Some of these helping hands reside in very deep pockets; pockets that belong to influential people connected to other influential people such as policymakers and of course our good friends in the media. Examining some of these connections ought to cast serious doubt on the illusion that the rise of Greta Tintin Eleonora Ernman Thunberg (for that is her full name) was a spontaneous school protest that morphed into a worldwide school strike.[1]

But before that, let us do a little thought experiment. I like thought experiments. If I do a thought experiment it means at least one piece of thinking was done that day.

Imagine you had a teenage son or daughter who was passionate about an issue where a burning injustice was being suffered in the world. For the sake of argument, let us choose child slave labour in Africa. Ten-year-old kids being sent to work into mines in the Congo, that kind of thing. Imagine your boy or girl got upset by this and was moved to take action. One day, he or she sat outside Parliament with a placard stating: 'Strike for child slave labour' and posted an image of the lone protest to Twitter and Instagram.

Ask yourself what are the chances that other children would voluntarily join the one child protest the very next day? You have to admit it is not the most enthralling prospect for a teenager. Is it even 50-50 that one single person would show up to join in? Would you be disappointed if nobody showed up or would you consider that to be the actual likelihood?

It might happen were the image to be shared hundreds of times on social media, so ask yourself, what are the chances of that happening? A kid sitting alone outside Parliament with a placard isn't the most exciting image, even if it is your child. How many likes and shares would your child's image attract? In the official version of the Greta Thunberg story, she canvassed her schoolmates in person beforehand and none of them were interested, yet after her first protest and social media pictures, interest magically flourished. Perhaps she is just not very persuasive in person.

Now, further imagine your local paper covers the story – *The Peckham Gazette* or whatever. Perhaps one of their journalists was just passing that day and considered it a newsworthy item. Is this even 10% likely? When you get to multiplying these probabilities out, it looks like we are going to end up with a staggeringly small number.

But let us carry on. A huge stroke of fortune benefits us. A very prominent child slave labour activist and entrepreneur sees this news in *The Peckham Gazette* and posts the story to all his followers on social media. (This is one of the 'lucky breaks' that Greta benefitted from. A man called Ingmar Rentzhog turned up at Greta's protest after seeing the story in the local news.[2] He was a businessman and environmentalist and he promoted the story to his social media followers.)

Then a prominent banker who just happens to work in the same field with 200,000 Twitter followers tweets your story. Just imagine! Of course, this is not an impossible turn of events. These things *can* go viral but what are the odds? Now imagine that within one week your child's cause is being taken up by media in France, Spain and the US, even though the story of a child with a placard outside Parliament is a little mundane.

The story appears on CNN and BBC, Russia Today, Al Jazeera. And so on. And on. And on. Within a year your offspring is shaking hands with Barack Obama, giving talks to the UN and enjoys the honour of being *TIME* person of the year (a rather dubious honour when you consider some of their former winners) with two consecutive nominations to follow for a Nobel Peace Prize (see previous comment).

What are the chances? Yes, that's right – the chances are zero, unless powerful people wanted it to happen. But, if that was the case, they would arrange their campaign with a child of their choosing. So, I'm afraid as wonderful as your child is and as noble as their cause, the chances are still zero.

It is so patently absurd on its face. When you see the people and the organisations who are involved with Greta, the only conclusion one can draw is that it is a setup. Some people planned the whole thing, just as if scripts were written. The whole thing was stage-managed and a lot of people were conned.

We have all heard the saying, 'The revolution will not be televised' so let us put it another way. If the revolution *is* being televised it is not the revolution. So what really happened? Let us return to the entrepreneur called Ingmar Rentzhog who helped boost Greta's cause.

Rentzhog is the founder of the Swedish consultancy firm, Laika, which provides communication services to the financial industry. He is heavily involved in the climate change business and is linked to Al Gore who has a track record of making vast sums of money from climate change initiatives. Rentzhog is a member of Al Gore's 'Climate Reality Organization Leaders' and was trained by Al Gore's organisation in March 2017 in Denver and again in June 2018, in Berlin. In May 2018, Rentzhog was appointed chairman of the think tank Global Utmaning, (Global Challenge) and since August 2018 he has also served on the board of FundedByMe, a crowdfunding platform. Last but by no means least, Ingmar Rentzhog is the founder and CEO of a Swedish company that runs a social network called 'We Don't Have Time'. As you can probably tell by the name, this business promotes the climate change agenda and promotes the specific message of the same name.[3]

This is the tweet Rentzhog posted via his 'We Don't Have Time' Twitter account that set the wheels in motion.

We Don't Have Time ●
@WeDontHaveTime0

Follow

One 15 year old girl in front of the Swedish parliament is striking from School until Election Day in 3 weeks

Imagine how lonely she must feel in this picture. People where just walking by. Continuing with the business as usual thing. But the truth is. We can't and she knows it!

5:40 PM - 20 Aug 2018

16 Retweets 42 Likes

👤 Greta Thunberg, Jamie Margolin, Zero Hour and 2 others

💬 1   🔁 16   ♡ 42   ✉

Before he promoted Greta's protest in August 2018, Rentzhog attended a climate event with Greta's mother Malena Ernman in May 2018. Later, Renzthog would fall out with the Thunbergs over allegations that he was making money from Greta's name; even the best-laid plans didn't run like clockwork. The key point is that Rentzhog was already familiar with at least one of the Thunbergs a full three months before Greta's protests began, so his 'lonely girl' tweet doesn't look quite so spontaneous. The name of his social network 'We Don't Have Time' has a familiar ring when you compare it to messages like 'our house is on fire' and the general notion that we need to act quickly because all kinds of apocalyptic brown stuff is going to hit the fan in 12 years if the global average temperature rises by 1.5 degrees (an opinion asserted as if it were some axiomatic truth).

The climate change issue is only one string to Rentzhog's bow. First and foremost, he is a businessman and a communications specialist. So when you hear Greta say 'our house is on fire' one has to wonder, did she think of that slogan herself or was it perhaps somebody else?

'We Don't Have Time' has a Founding Manifesto which was written in April 2018, a month before the May 2018 climate meeting with Greta's mother. Again this suggests that a campaign was set in motion long before Greta's protests started. Was Ingmar Rentzhog, the communications specialist who learned to promote the message of climate fear from the master Al Gore himself, hired to create a viral marketing campaign? If so, who hired him?

The Founding Manifesto begins with the words: 'We are in serious danger'. Other prominent headings are 'Mobilisation', 'Awareness', 'We Don't Have Time' (again) and 'You can be the Change'. Those communication skills again. It also includes the ominous warning that 'climate change is killing us'.[4] Small wonder Greta often looks miserable when she is surrounded by adults who promote this kind of rhetoric.

There is so much more to be said about the public relations nature of this campaign. Here are some selected extracts from a blog post written by Cory Morningstar, a genuine environmental activist and to my mind the best researcher bar none on this particular subject. It catalogues part of the timeline surrounding the rise of Greta Thunberg.[5]

2017: World Economic Forum founder Klaus Schwab: 'Capitalism is in crisis'

2018: A teleconference led by a 350.org/Fossil Free representative with Climate Reality Project (Al Gore's NGO) proposes a large climate march. Greta Thunberg partakes in this call as well as others that transpire. The idea of a strike is presented. Thunberg is receptive.

May 2018: Ingmar Rentzhog, founder and CEO of We Don't Have Time, is featured at a climate event with Greta's mother Malena Ernman.

June 2018: Greta Thunberg's social media accounts are created.

Summer/Fall 2018: The Green New Deal (promoted by UN in 2009) is resurrected.

July 2018: The Climate Group, co-founder of We Mean Business, promotes This Is Zero Hour climate strikes in the US utilizing the hashtag #WeDontHave Time ['Join the youth revolution!']

20 August 2018: Greta sits on a sidewalk with a sign. Rentzhog discovers 'the lonely girl'. We Don't Have Time, partner of The Climate Reality Project, and Global Utmaning (Global Challenge) are interconnected by board relationships.

20 August 2018: On the first day of the strike, the third person to respond to 'the lonely girl' plight on Twitter is We Mean Business co-founder Callum Grieve. He adds the hashtag #WeDontHaveTime and tags five additional accounts: The Climate Museum, Youth Climate March LA, This is Zero Hour Ft Lauderdale, Greenpeace International and the UNFCCC, the 'official Twitter account of UN Climate Change'.

These details paint a very different picture to the one portrayed by the media. The third person to respond to the 'lonely girl' tweet was Callum Grieve, who has an impressive CV.[5]

"Grieve is the co-founder and director of Counter Culture, a brand development firm specializing in behavioural change campaigns and storytelling. He also created Climate Week NYC for The Climate Group. Grieve has coordinated high-level climate change communications campaigns and interventions for the United Nations, the World Bank Group, and several Fortune 500 companies."

Like Ingmar Rentzhog he possesses a particular set of skills. Callum Grieve is a communications specialist with a long list of important business contacts. Of course, the UN and World Bank are enormously powerful institutions that are intimately involved in the net-zero carbon agenda. Perhaps the most intriguing comment is that Grieve co-founded a 'brand development firm specializing in *behavioural change campaigns and storytelling*'. (Emphasis added)

So many questions spring to mind. Why does this message require PR experts and why the sleight of hand? Whose behaviour requires changing? Why are so many billionaires and financial companies involved? And what is the 'story' being told? Who is hiring these people and how much money is changing hands to promote this message?

The name of the company, 'We Mean Business', offers a clue. We Mean Business is a coalition of organisations that work with businesses and investors ostensibly committed to 'bold climate action'. Callum Grieve was its communications director in 2018. It seems that they really do mean business, representing 1,965 companies responsible for $24.8 trillion of assets.[6]

Climate change is a cold calculated business opportunity for many of the companies professing to care about the environment. Literally, trillions of dollars are going to change hands in the name of 'net carbon zero' so it is no surprise we run into slick PR men when we scratch the surface. The blog post continues:[5]

"Grieve also manages the Every Breath Matters campaign founded by Christiana Figueres, the former UNFCCC Executive Secretary credited with the Paris Agreement. Every Breath Matters 'champions' include Leonardo DiCaprio and Greta Thunberg."

Christina Figueres is an incredibly important individual in the climate change arena, as we will see very shortly. The UNFCCC is the UN Framework Convention on Climate Change, a multilateral agreement created as a result of the 1992 Rio Earth Summit. The Kyoto Protocol and Paris Agreements were born from the UNFCC.

Returning to Cory Morningstar's article, we can see distinct coordination between the climate story being told, so-called philanthropy and big finance.[5]

- **20 August 2018:** Also on the first day of the strike – the 'lonely girl' plight is shared by Sasja Beslik, international financial expert (WEF), head of Sustainable Finance, Nordea Bank.
- **Fall 2018:** New Deal for Nature and Voice For The Planet campaigns commence. Exploiting an increasingly anxious citizenry, utilizing emotive images and language, these campaigns are in fact, not to 'save nature', rather, they are to monetize nature, global in scale.
- **1 September 2018:** Only 12 days after her first day sitting on a sidewalk, Greta is featured in *The Guardian*.
- **September 2018:** The largest-ever philanthropic investment to combat climate change is announced by ClimateWorks, the largest recipient of climate philanthropy in the world.
- **26 September 2018:** Thunberg appears at a seminar organized by The Climate Reality Project and Global Utmaning (Thunberg's father denies any relationship or affiliation with Global Unmanning).
- **26 September 2018:** The Climate Finance Partnership – a vehicle for blended finance – is unveiled at the One Planet Summit.
- **31 October 2018:** Launch of XR (Extinction Rebellion) global expansion is highlighted by *The Guardian* and endorsed by an array of liberal celebrity signatories.

Cory Morningstar draws countless connections between PR gurus like Rentzhog, Grieve and Al Gore (for that is what he is, a PR guru for the climate change story), Non-Governmental Organisations (NGOs), national governments, supranational bodies such as the

World Bank, the UN, the International Monetary Fund (IMF) and, of course, businesses from Wall Street and the City of London.

It is all a far cry from the lonely girl on the pavement.

When one has digested the sheer cynicism of this public relations operation, the only thing amazing about the rise of Greta Thunberg is how many people were taken in by it. That must include Greta Thunberg herself. She could have had no idea of the scale of the global project that used her as its figurehead.

Greta Thunberg pictured with Barack Obama, under whose administration oil production increased for 7 out of 8 years and whose campaign of drone warfare killed thousands of children.

1.2

The Financialisation of Nature

Note that the very first entry from above, dated 2017, is where the head of the World Economic Forum, (WEF) Klaus Schwab is quoted. It is imperative to understand just how important the role of the WEF is when it comes to climate change issues. By extension, the future direction of the world economy and indeed the well-being of the entire human race (or not, as we shall see is more likely to be the case) appears to lie in the hands of this body. Also, note that the 'Green New Deal' had already been promoted by the UN as far back as 2009. These plans are not new. Another incredibly important entry in the list above is this one:

'World Economic Forum UN Partnership Effective June 13, 2019.'

The UN is supposed to act on behalf of all member countries, which means it is supposed to act on behalf of all the citizens of the world. Yet, in June 2019, the UN formed a partnership with the WEF to very little fanfare. The UN, which notionally represents the whole world, teamed up with the Davos crowd – the billionaires, the 0.001% – and the vast majority of the world did not even notice. One hopes that the partnership is not a takeover but when there is a relationship between billionaires and bureaucrats, one can guess who the senior partner is.

In this connection, a big hint was dropped by Christina Figueres, the highly ranked UN official mentioned earlier. Between 2010 and 2016, Christina Figueres was the Executive Secretary of the UN's Framework Convention on Climate Change (UNFCCC). This position is a direct appointment of the UN Secretary-General, who at the time was Ban Ki Moon. Christina Figueres is therefore not a minor player and she was not speaking off the record.

On 3 February 2015, Figueres casually made a startling admission at a press conference in Brussels. She was speaking about a proposed international climate agreement that environmentalists hoped would be finalised in Paris later in 2015 – the 'Paris Agreement' – which was indeed signed on 22 April 2016.

Figueres said,[7] "This is the first time in the history of mankind that we are setting ourselves the task of intentionally, within a defined period of time, to change the economic development model that has been reigning for at least 150 years, since the Industrial Revolution."

Changing the economic model? The economic model Christina Figueres was referring to is clearly and unambiguously the economic model known as capitalism. Wasn't the Paris Agreement supposed to be about countries agreeing to be more environmentally friendly?

She continued, "This is probably the most difficult task we have ever given ourselves, which is to intentionally transform the economic development model for the first time in human history."

This was an astonishing declaration and one being admitted to in public.

Changing the 'economic development model' of the last 150 years can only possibly mean replacing capitalism (or if you prefer, free enterprise). The word 'intentionally' leaves no room for ambiguity and she said it twice. Figueres also said this was to be done 'within a defined period of time', which almost certainly refers to Agenda 2030, a UN plan that was rebranded as the UN's 17 Sustainable Development Goals. It appears there is a deadline to replace capitalism by 2030.

These comments really ought to have made a bigger splash than they did. Here was the chief guardian of the Paris Agreement admitting that the UN wanted to completely change the economic model of capitalism, to do so deliberately and to do it within a certain timeframe. Figueres was talking about changing everyone's lives. Literally, everyone's. Do citizens have a say in the matter before some UN czar blithely states that capitalism is being consigned to the dustbin? And what will be replacing it?

Figueres mentioned 'the task we have given ourselves' but who appointed the 'we' she refers to, namely herself and her UN colleagues? Should it not be up to individual countries to decide what economic model they adopt? What gives the UN the right to try to change the entire world so profoundly over the heads of nation-states? It's not as if any of the UN staff have a mandate to carry this out, even if they were elected, which none of them are. That also applies to their new partners at the WEF who are helping them to roll this out. Rest assured the members of the WEF *are* there to help roll out this new model. In fact, they are the likely drivers of the change.

The WEF might have only formed an official partnership with the UN in the middle of 2019, but the planners and schemers at the WEF have been advocating changes to the world's economic model for a very long time. Talk of 'stakeholder capitalism' is now decades old and references to an 'economic reset' can be found on the WEF website dating back to at least 2014.

Recently the WEF has been very busy. The head of the WEF, Klaus Schwab, has authored or co-authored four books whose titles provide clues to their intentions.

January 2016 – *The Fourth Industrial Revolution*
January 2018 – *Shaping the Fourth Industrial Revolution*
July 2020 - COVID-19 – *The Great Reset*
January 2021 – *Stakeholder Capitalism*

Klaus Schwab's July 2020 book *COVID-19 – The Great Reset* was published shortly after the WEF's lavish presentation of the same name on 3 June 2020 featuring Prince Charles and other dignitaries. The great and the good came together to announce that, in their infinite wisdom, a complete restructuring of the world's economy – and in fact, the world itself – was somehow required, a need apparently brought about due to COVID-19.

In the WEF's own words, "The Great Reset is a new initiative from the World Economic Forum and HRH the Prince of Wales to guide decision-makers on the path to a more resilient, sustainable world beyond coronavirus."

One might raise an eyebrow at Schwab's description of the initiative as 'new' given the WEF's repeated references to the word 'reset' over recent years. As to 'resilient' and 'sustainable', these are adjectives one sees again and again when reading WEF documents and even more so in the blurb emanating from the UN. These words are rarely if ever, defined but repeating them signals a particular set of values to certain people, shall we say. Keep your eye out when perusing future WEF and UN documents and you will be amazed at just how often these words crop up. (See also 'vibrant', 'diverse' and 'empowering'.)

Unveiling the Great Reset presentation, Klaus Schwab said on 3 June 2020:

"The pandemic represents a rare but narrow window of opportunity to reflect, reimagine, and reset our world", a quote that can still be found on the Great Reset page of the WEF's website.

An almost identical quote can be found in the 'Conclusions' section of Schwab's book *COVID-19 – The Great Reset*. A publication date of July 2020 implies that Klaus Schwab believed the world needed to be reset some time before and possibly several months before July 2020. How could anyone have possibly known what the effects of COVID-19 would be this early and have deduced that an economic reset was required?

'Reset' is a word that has been increasingly drip-fed into the public consciousness since around 2014, through white papers, news organisations and the instigators themselves. Anybody who is tempted to use COVID-19 as an excuse to reboot capitalism might as well call it the 'Great Non-Sequitur'. There is no good reason why COVID-19, or any pandemic for that matter, should lead to wholesale economic or social change. One does not follow from the other. One gets to thinking that any excuse will do for these people.

What also defies logic is that the people replacing capitalism are the 0.001%, the uber capitalists themselves, the Davos crowd. These are the very people who did best out of capitalism, who have a proven track record of earning colossal sums of money while ransacking the planet, yet who now believe capitalism is old hat and that we need to be better stewards of the earth. That ought to strike any thinking person as being slightly odd.

One thing does start to make sense though. A hard lockdown across all the major economic centres of the world for a not very threatening respiratory illness would speed up the end of capitalism, would it not?

The UN and WEF are using the climate 'emergency' to hasten the replacement of capitalism and the Green New Deal is a central pillar of the replacement exercise. They are also opportunistically using COVID-19 to argue for a new economic model. These cannot be labelled 'conspiracy theories' because they have admitted it in their press conferences and books.

Very much like the 'Fourth Industrial Revolution', the 'Reset' is something we will be hearing a lot about in the coming years. Expect the mainstream media coverage of both the 'Reset' and the '4INR' to be unanimously positive. They will be presented as *necessary* changes. They will be presented as inevitable changes, just like the move to a cashless society is presented as inevitable and as a necessary improvement to our lives.

I wonder how many people realise what replacing capitalism actually means. Some people might think ending capitalism is a good idea because of the inequality and poverty it has inflicted on so many, not to mention pollution. While this is understandable, people need to be careful what they wish for. Think carefully about what we will be giving up.

The hallmarks of capitalism are private property rights and personal freedom. You can bet these are the first things on the chopping block if capitalism is replaced. This is why the term 'free enterprise' is more preferable to 'capitalism'. For all the very real criticisms of capitalism, free enterprise really did allow much of the world to flourish for a very long time.

Let us not believe for one second that there will be no inequality if the UN-WEF plan of replacing capitalism comes to pass. It is more likely to be a two-tiered system with an absolutely, minuscule number of super elites and then one giant roughly 'equal' strata of humanity beneath where billions of us will live at a very low level in smart cities with energy rationed on smart meters, all heavily monitored, but yes, broadly equal. Equality of the giant lower strata will not be reached by raising African standards of living towards western standards but precisely the reverse.

When the WEF is issuing speeches and articles written by the Danish MP, Ida Auken, asking the questions: "What would a world without property look like?" and "Why would you want to own your own phone?" we should be very wary.[8] Anyone who tells you that you do not need personal property has an agenda. Ida Auken's article, titled *Welcome to 2030. I own nothing, have no privacy, and life has never been better* has to be read to be believed. It betrays an astonishing level of delusion on the part of the WEF that the public will accept the world they wish to steer us into, a world with no possessions or even privacy.

It is unlikely the elites will be giving up their yachts any time soon and I do not relish the prospect of not owning my own bicycle, my own washing machine or even my own clothes. One only needs to look at the WEF's website to see their vision of the future and for all their word salads and messages of positivity, it all looks rather dystopian.

Capitalism did lift billions of people into a decent standard of living, but it was bastardised into the crony capitalism we see today where vested interests always win out. Political lobbying – legalised bribery that ought to be outlawed – and the buying of influence where the biggest corporations literally draft legislation, has corrupted the system so badly the world has been left in the position we are today. Ask yourself who the biggest exponents of this mafia economics are. One might conclude that the biggest culprits are the banks, pharmaceutical companies and technology giants. Cross-check the names of the corporations you arrived at against a list of the WEF members. You are guaranteed to find your chosen names there.

Understandably, people hate capitalism but things can get a lot worse.

By the way, it would be a mistake to assume that just because the opposite of capitalism is communism that these people are communists. There is not a lot of evidence of them doing much wealth sharing and I wouldn't be fooled by their 'philanthropy' either. Certain WEF members will be making money hand over fist through the Green New Deal and Fourth Industrial Revolution. Judging by their actions and not what they say, they haven't given up on the idea of accumulating wealth.

No, the reset will not usher in communism in place of capitalism, although it will certainly feel like communism for billions of us. It is a different system they are thinking of and its name is 'Technocracy'. To those on the die-hard left and the communist end of the political spectrum who are cheerleading the death of capitalism, consider this. The fact that it is the Davos crowd rolling out the new model ought to fill you with questions, if not concern, if not outright dread.

As you can see from the share prices of Amazon, Apple, Walmart, Microsoft and others, *that* kind of capitalism still works OK for some. Better than ever in fact, even during a lockdown.

You are probably wondering where Greta Thunberg fits into all this.

Greta is the pin-up child for the Green New Deal and the idea of using Greta is to attract as many young people as possible to join the cause. Remember Callum Grieve and his 'behavioural change' campaign with 'storytelling'? The millions of kids who joined the school strikes for climate will remember those days off school and they will feel personally invested in the cause and all its goodness. Saving the environment is a noble aim indeed. But this is not what the Green New Deal is all about. The Green New Deal is a means to justify replacing capitalism. The Green New Deal is a front for the UN and the WEF's plans.

It is so important that it needs repeating; as counterintuitive as it sounds, the Green New Deal is not about protecting the environment. Using climate concerns to affect economic policy is not a new phenomenon and this has been blatantly admitted. In

November 2010, Dr Ottmar Edenhofer, who was the co-chair of the Intergovernmental Panel on Climate Change (IPCC) Working Group 3 said,[9] "We redistribute de facto the world's wealth by climate policy... One has to free oneself from the illusion that international climate policy is environmental policy. This has almost nothing to do with environmental policy anymore."

Anyone arguing that this quote was taken out of context would have a hard task explaining how.

Those kids who skipped school to save the climate once upon a time, who nagged Mum and Dad to fly less and so forth, will equate the Green New Deal with the new economic model and believe in its righteousness and justice. By association, they will be cheering for the end of capitalism. But they are being misled. Through PR campaigns and appeals to emotion, they have been sold a vision of the future that is not going to happen. The Green New Deal is about economics and the financialisation of nature. The winners will be those companies rewarded with government contracts for 'renewable' schemes such as wind, solar and carbon capture systems of yet unproven value. The losers will be almost everyone else – forced to pay higher energy prices, being unable to travel, going cold in winter and deprived of carrying out regular activities.

What the 'financialisation of nature' means is that all 'ecosystem services' that provide benefit to humans will be assigned a monetary value in the new system. Nature's wonders will be standardised, objectified and commoditised, to be traded on financial markets: sliced, diced and packaged like mortgage-backed securities. Think carbon credit trading on steroids. The 'externalities' of industrial output such as pollution and emissions will be assigned negative values, with ecologically destructive companies paying for credits – in effect a licence to pollute – while they claim with a straight face to be carbon neutral or even environmentally friendly.

Banks and hedge funds will be the arbiters of value for these products. For these players, capitalism will be thriving, regardless of any talk about capitalism being replaced. For the rest of us, there will be Payments for Ecosystem Services (PES) and, in the future, carbon rationing is likely to be the order of the day. Ultimately, all-natural resources will be up for grabs: fuels, forests, reservoirs, livestock and even humans.

That's right – even us humans will be treated as resources one day as if we were mere inputs on a production line. If you find this difficult to believe, try an internet search for 'social capital bonds' or read a white paper on 'human capital performance bonds'. These financial products were conceptualised long ago and already exist, ready to be rolled out in size when the time comes. When they are scaled up it will appear as if they emerged from nowhere. Educated people will wonder why they had never heard about these ghastly products. This is the dark side of economics. Their creators were never going to shout about these inventions from the rooftops, even less so the media. The media's job is to get the kids focussing on nebulous concepts like helping the earth to breathe and being good global citizens.

The average child will not understand the financialisation of nature, explained in jaw-dropping detail by Cory Morningstar or what the new economic system will entail. Mentored into the philosophy that the ends justify the means, they will believe any policy unveiled under a 'Green' banner is worth it to save the planet and that is all that matters. Of course, the apocalyptic '12 years to save the earth' deadline takes us right to 2030.

Children are generally more suggestible than adults, more likely to follow a movement with noble aims and the threat of a doomsday scenario such as a mass extinction. By definition, they have not lived through the 1970s, 80s and 90s where the same old climate change horror stories were recycled over and over without ever coming true.

The older generations have heard it all before. They also have ingrained habits and some people are just not for changing. A fifty-year-old who has been driving for thirty years probably isn't going to like the idea of expensive electric cars that can only do short journeys. The idea of driverless cars will appeal even less. These people are far harder to convince than children and it is not even worth trying to persuade them. Eventually, that generation will die off and the children who are now adults will still be believers. This is why the people who engineer these big plans target the youth.

And that is where Greta fits in.

1.3

Environmental Carnage

There are aspects of the reboot of capitalism that should alarm both businesses and individuals. Whilst it is unsurprising that politicians muscle in on photo opportunities with media darlings such as Greta Thunberg, the sight of central bankers like Christine Lagarde and Mark Carney in Greta's orbit requires an explanation. These two individuals have interesting backgrounds.

Christine Lagarde is a lawyer by profession and is currently the head of the European Central Bank (ECB). Before this, she was the Managing Director of the IMF. She was convicted of criminal negligence in 2016 for misuse of public funds when she was French Finance Minister in 2008, but escaped a prison sentence and kept her IMF job.[10]

Lagarde was appointed as Chair and Managing Director of the IMF after the previous incumbent, Dominique Strauss-Khan, resigned due to being prosecuted for an alleged sex assault on a hotel chambermaid. The charges were eventually dropped, leading to well-founded suspicions that the French banker had been set up in a sting operation, but by then he had already been replaced.[11] Lagarde's later appointment to the governor of the ECB was despite her criminal conviction, so it is safe to conclude that certain powerful people want her to hold these top positions. It is not as if there aren't legions of other qualified candidates without a criminal record.

On 25 January 2021, Lagarde gave a speech in Frankfurt where she sent a strong signal about the role of central banks in addressing climate change.[12] She was clear that central banks *did* have a role to play with respect to the environment, saying, "The fact that we are not in the driving seat does not mean that we can simply ignore climate change or that we do not play a role in combating it."

Whilst acknowledging the risk of 'mission creep' in her speech, if history is any guide, she was only paying lip service to it. In time-honoured central banking tradition, the central banks' missions will not just be creeping but stampeding.

The original role of a central bank was to provide a sound currency and that was it. This role was later expanded to 'lender of last resort', the idea being to stave off financial crises by providing funds to banks at interest in exchange for high-quality collateral. Today, central banks lend money at zero interest in virtually unlimited quantities to favoured borrowers in return for low-quality collateral and often no collateral at all. Next came the setting of interest rates and mandates to affect the employment level, liquidity and almost every area of the economy. Today, central banks buy up bonds and mortgage-backed securities by the bucket load and even equities as well. The Japanese central bank owns around half the country's government bond market and is the biggest holder of Japanese equities. The Swiss National Bank prints Swiss francs in any limit it chooses and sells them for other currencies. Being the sovereign issuer of a strong currency which it does not mind weakening from time to time, it is quite able to do this. But when the Swiss National Bank switches currency printed out of thin air into US dollars to buy billions of dollars' worth of shares in Apple, Google, Facebook and General Electric – well this is a far cry from lender of last resort.

Not content with this ever-growing pile of assets, central banks have now set their targets on climate change. Let us see this for what it is – another excuse to control the economy, and by control – read plunder. When Christine Lagarde politely informs listeners that the European Central Bank is considering playing a role in combating climate change, take it as read; it is a done deal and the decision was made some time ago.

Even if one believed their motives were pure and the goals laudable, there must be a more appropriate body than a central bank to take on this responsibility. Central banks trying to influence climate change indeed. Have you ever heard anything so ridiculous? What next – using their regulatory powers to ensure racial equality or to fight sexism? Will mission creep extend to their being the arbiters of misgendering pronouns after painting the ECB green – the 'gender of last resort'?

Christine Lagarde has another influential role. She is a member of the board of trustees of the WEF. Other notable individuals on the WEF board of trustees are Alibaba's Jack Ma, Blackrock's Larry Fink, Al Gore and the aforementioned, Mark Carney.

Mark Carney was the head of the Bank of Canada before becoming head of the UK's Bank of England in 2013. He held an arguably even more influential role than both of these central banking jobs when he was the Chairman of the Financial Stability Board from 2011 to 2018, a body that is part of the Bank for International Settlements (BIS).

The Bank for International Settlements, based in Basel, Switzerland, is the apex of the world's financial system. All current 62 members of the BIS are either central banks or monetary authorities. Central banks' representatives meet in Basel to co-ordinate world monetary policy and crucially, this process is carried out away from the gaze of politicians. Carney also served as Chairman of the BIS Committee on the Global Financial System from July 2010 until January 2012.

Before becoming a central banker, Mark Carney worked for Goldman Sachs for 13 years. If you have ever wondered why the world's central banks and governments' finance ministries are full of Goldman Sachs alumni (Mario Draghi, Mark Carney, Hank Paulson, Timothy Geithner, etc), or why people with nine-figure fortunes would take administrative government positions for six-figure salaries, you are not alone. It is worth remembering the adage a wise man once uttered: one never retires from Goldman Sachs.

Mark Carney has other roles. He is a member of the little-known Group of Thirty, a peculiar think tank consisting of private bankers, central bankers and academics. The Group of Thirty was set up and financed at the behest of the Rockefeller Foundation in 1978. Despite the enormous potential for conflicts of interests – especially in light of the common employment history of central bankers just mentioned – the group holds closed-door meetings to discuss industry issues. The very existence of this group has been described as 'scandalous'.[13]

Mark Carney attended the annual meetings of the Bilderberg Group in 2011 and 2012. The secretive Bilderberg Group is a forum where business leaders and politicians meet in luxury hotels on an invite-only basis to plan... well who knows what they plan? The meetings are not minuted, at least not publicly, although vague agendas are posted on the Bilderberg website. Attendees never talk about what was discussed at meetings, at least not if they want to be invited back. The lack of transparency is ironic given that they make a point of inviting starry-eyed journalists and academics, especially from *The Financial Times* and *The Economist*, who are under strict orders to report absolutely nothing of any consequence. Meetings are guarded by a heavy police presence.

Bilderberg meetings have drawn a lot of criticism and quite rightly so. If an executive of Shell Oil wants to meet privately with an executive from Barclays at their own expense and keep the content of that meeting secret, there is nothing wrong with that. They are private individuals. However, when the Finance Minister of a large European country is invited, or a Deputy Prime Minister or, interestingly, the UK's head of media regulator OFCOM, then the secretive nature of these meetings does become an issue. These people are supposed to be public servants.

Mark Carney has never been elected to any of these positions but with his many hats, he is extremely well connected. Now he has a new hat.

On 1 December 2019, UN Secretary-General, António Guterres, announced the appointment of Mark Carney as UN Special Envoy for Climate Action and Finance.[14] Carney would be paid a token $1 a year for the role and he would replace Michael Bloomberg (net worth approximately $50 billion) who stepped down from the role citing a 2020 Presidential bid that did not materialise.

Carney said the new role would provide, "… a platform to bring the risks from climate change and the opportunities from the transition to a net-zero economy into the heart of financial decision-making."

Quite. Well, he wasn't brought in as a scientist or for his love of polar bears.

He said, "Investing for a net-zero world must go mainstream. The Bank of England, the UK government and the UK financial sector can play leading roles in making these imperatives happen."

This statement deserves closer scrutiny. Carney believes net-zero investing *'must* go mainstream' and is an *'imperative'*. This language shows that he is serious. He made an even more assertive declaration in 2019, showing just how intent he is on making this a reality. On 1 August 2019, *The Daily Mail* reported Mark Carney as saying:[15]

"Companies that don't adapt, including companies in the financial system, will go bankrupt without question."

The UK's *Guardian* newspaper reported similar comments from Carney on 13 October 2019:[16]

"On Tuesday, Carney told big corporations they had two years to agree on rules for reporting climate risks before global regulators devised their own and made them compulsory."

Mark Carney said, "There will be industries, sectors and firms that do very well during this process because they will be part of the solution. But there will also be ones that lag behind and they will be punished."

Sounds a little bit like a threat, doesn't it?

What Mark Carney means is that companies that do not toe the line with emissions planning, regulatory reporting and everything else mandated by net-zero rules, can look forward to credit rating downgrades which will prevent them from issuing bonds to raise finance. Central banks could even require that banks lend money to wind and solar firms and deny credit to firms deemed to be disobedient, thus starving them of cash. (This is the very risk that wise men have warned against for centuries. The Founding Fathers knew all about this in the 1770s. Americans abolished two central banks in the 19th Century because of these fears. Yet here we are today.)

Like Callum Grieve and Ingmar Rentzhog, Carney means business. Only in this case, he means punishing businesses and even bankrupting businesses that do not obey the rules. This is what he means by putting 'the transition to a net-zero economy into the heart of financial decision-making'. This is the financialisation of nature writ large and Mark Carney is telling businesses in 12 point red font: comply or die. Small wonder people call him Mark Carnage.

Larry Fink agrees with Mark Carney. Fink is the CEO of the world's largest asset manager Blackrock Inc. Blackrock Inc has become perhaps the world's most influential finance company – even more so than Goldman Sachs – holding great sway with politicians, central banks and the world's CEOs. An asset manager!

In 2021, Blackrock was managing in excess of $9 trillion of assets, a phenomenal trove that included stakes in all the world's largest listed companies. That kind of share ownership means an awful lot of voting rights are either exercised or influenced by Blackrock Inc. Larry Fink's opinions on corporate governance really do count, and Fink has got the green bug too – pretty much a given seeing as he is a member of the board of trustees of the WEF. In his 2021 letter to CEOs, Fink wrote:[17]

"We have long believed that our clients, as shareholders in your company, will benefit if you can create enduring, sustainable value for all of your stakeholders."

Fink added that 'climate risk is investment risk' and that 'the creation of sustainable index investments has enabled a massive acceleration of capital towards companies better prepared to address climate risk.'

Yes, cash will flow to the 'good guys'. He drops a hint as to how CEOs might consider their future business strategies:

"As more and more investors choose to tilt their investments towards sustainability-focused companies, the tectonic shift we are seeing will accelerate further. And, because this will have such a dramatic impact on how capital is allocated, every management team and board will need to consider how this will impact their company's stock."

Fink's message is more carrot than stick, but it boils down to the same thing as Mark Carney was saying: comply or die. Uncle Larry was making CEOs an offer they couldn't refuse.

When someone like Mark Carney predicts that businesses 'will go bankrupt without question' you can take it to the bank, as it were. Mark Carney's strident talk of punishing and bankrupting firms for not being green enough might seem odd coming from a central banker who had not even started his $1 a year climate job at the UN. However, Mark Carney was already involved in the climate change arena from at least 2015.

Back in 2015, as chair of the Financial Stability Board – the important role previously mentioned – Mark Carney established the 'Task Force for Climate Disclosure (TFCD)' in response to a request from the G20 to better understand the financial risks posed by climate change. He appointed Michael Bloomberg (net worth approximately $50 billion) as its chair.

'Climate disclosure' refers to financial information disclosed in company reports concerning climate change. For instance, an insurance company that is running financial risks from claims on adverse weather events could make a disclosure in its financial statements and annual report. Much of this is voluntary but the TFCD would like to change this and introduce hard reporting requirements with global standards. It is worth noting at this stage that TFCD has no official power to do any of this. It is just a body set up by Mark Carney working for the Financial Stability Board (FSB). The FSB is part of the Bank for International Settlements, which is itself a private company whose officials are not elected.

The chair of the TFCD, Michael Bloomberg (net worth approximately $50 billion – I think you've got the message now) is a businessman and a private citizen.

In their opinion, the disclosure of climate-related information by companies is necessary to allow investors to shift investments to low-carbon economies and avoid financial stability risks. While it is true attitudes are changing in this regard with more investors wishing to 'go green', largely due to the worldwide green media onslaught (see Greta Thunberg and friends) it rather presumes that *everyone* prefers to invest in greener technologies. If that was true there would not have been a pollution problem in the first place.

Businesses naturally gravitate to lower-cost models, not expensive green ones. The green revolution is not a natural phenomenon but a manufactured one. That manufacturing effort now includes overt coercion, as articulated by the likes of Mark Carney and Larry Fink.

Carney and Bloomberg's TFCD have been working hard to implement global reporting standards. "… we are also working with the global standard setters to drive a comprehensive approach. So, all of these approaches should start to move together," Carney declared in May 2020.[18]

It is against these standards that companies will be measured and measurement will determine if they survive or not, recalling the stern rhetoric from Carney in 2019.

* * *

As an aside: this is how globalism works. Unelected persons establish focus groups and bodies like the TFCD. Friends from the business world and the 'right kind' of academics are hired to write white papers and issue guidance. Calls for 'harmonisation' are made, always for the 'greater good' and to iron out inconsistencies in the name of 'co-operation'. After all, only nasty people don't want to harmonise and co-operate. Before long, sovereign countries with supposedly independent judiciaries are all implementing the same rules. While these rules exist on their own country's statute they were handed down to them by unelected elites like Michael Bloomberg.

The Guardian reported on the TFCD on 8 October 2019:[19]

"The former New York mayor has signed up a range of banks, asset managers, pension funds and insurers that control balance sheets totalling $120 tn (£98 tn) [trillion] to the project. He says four-fifths of the top 1,100 companies operating from the G20 countries are disclosing climate-related financial risks in line with some of the TFCD recommendations.

The aim is to develop a series of rules governing how companies report the effect of warming temperatures on their businesses alongside data showing the contribution of their activities to the problem.

With trillions of dollars needed to build climate-friendly infrastructure, investors needed to know which companies to back and which to shun, Carney said."

One would have thought it was simpler to shun/punish/defund/bankrupt the businesses that were causing the pollution rather than the ones who didn't make the correct disclosures.

We can think of the TFCD as the body designated to drive through the destruction of non-compliant businesses. Traditionally, it is the largest companies that can absorb increased regulatory costs and there is no reason to believe things will be any different here.

One particularly galling aspect of 'climate disclosure' was a comment made by Michael Bloomberg in July 2019 when presenting the 2019 TFCD report:

"I believe in the power of transparency to spur action on climate change *through market forces*."[20]

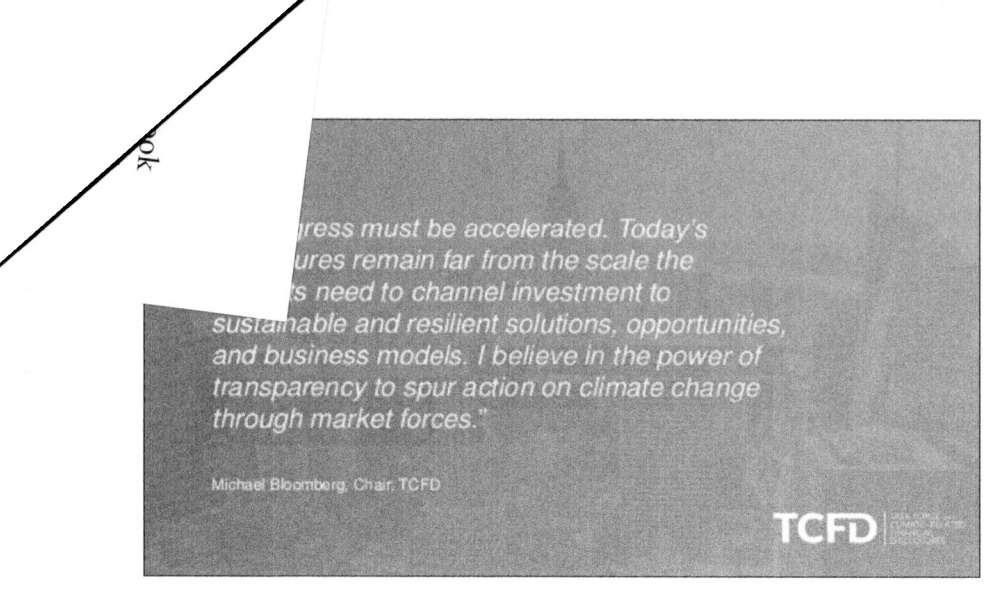

An insight into Michael Bloomberg's sense of humour.

Market forces? It is not market forces being applied here. It is the brute force of Mark Carney and his billionaire friends dictating what rules must be followed with threats of punishments and bankruptcy.

Punishment will be meted out via the banking system providing credit to companies that are declaring their efforts to move toward net-zero carbon emissions and withholding it to those that are deemed non-compliant.

It is closer to mafia tactics than market forces. Michael Bloomberg knows exactly what 'market forces' are and so does Mark Carney. One can only assume they are having a little joke at everyone's expense.

Let's recap some of the facts. Mark Carney works for the UN and he is on the board of trustees at the WEF. To those wondering if the UN-WEF partnership was a takeover and who was dancing to whose tune, in this case, they are the very same people.

Carney used to be the Chair of the Financial Stability Board at the Bank for International Settlements so he is on first name terms with all of the world's major central bankers. The same can be said for Christine Lagarde from her IMF days. She is also on the board of trustees of the WEF and now she is head of the European Central Bank.

They are in the perfect position to persuade central banks to pressure domestic banks to make life hard for businesses that aren't 'green' enough. Their mission dovetails nicely with the UN's declared objective of replacing capitalism. Capitalism relies on market forces, at least in theory. Kneecapping businesses for not following rules made up by Michael Bloomberg has nothing to do with 'market forces' and Carney is on record saying that the naughty firms will go bankrupt. They are deadly serious.

There was another comment from Mark Carney in the October 2019 *Guardian* article which is of interest:[19]

"Mark Carney also told the Guardian it was possible that the global transition needed to tackle the climate crisis could result in an abrupt financial collapse. He said the longer action to reverse emissions was delayed, the more the risk of collapse would grow."

This is a strange statement that looks like a complete non-sequitur at first glance. Carney appears to be saying that if businesses keep polluting, it risks an economic collapse even though the reverse has been true for 150 years. Fossil fuel consumption has driven economic growth with excellent correlation. He offers no explanation as to how this perceived climate crisis will cause a financial collapse but on closer inspection, this is not what Carney is saying. He is not claiming it is pollution or CO_2 emissions that will lead to abrupt financial collapse, but the '*global transition needed to tackle the climate crisis*'. Even assuming there was a climate crisis that required a global transition, he is admitting that the imposition of the transition rules will *cause* the collapse – the very rules he promotes in his UN role! This is a tacit admission that his actions could result in an abrupt financial collapse. At least he gets marks for honesty. A bit like Christina Figueres.

It is easy to predict a crash if one knows a crash is due. With a decade of zero rates and trillions having been printed via quantitative easing, we can rest assured that a crash is due. A crash is also a handy thing if one's task is to replace capitalism, which we can assume is the case given that Carney now works with the UN environmentalists whose stated goal is to replace capitalism. Those lockdowns must help as well.

Mark Carney referenced one other piece in the financial collapse puzzle in a speech on 23 August 2019 at the Jackson Hole Symposium, an annual meeting of the world's central bankers, finance ministers and economists. He dropped a huge hint as to the future reorganisation of the world's monetary system, stating in typically long-winded banker terms that the US dollar's days as reserve currency were coming to an end. *The Financial Times* reported:[21]

"Mark Carney, the Bank of England governor, has said that the world's reliance on the US dollar 'won't hold' and needs to be replaced by a new international monetary and financial system based on many more global currencies. In a speech at the annual Jackson Hole gathering of central bankers in the US, he called for the IMF to take charge of a new system of currencies, insuring emerging economies from destructive capital outflows in dollars and removing their need to hoard US currency. In the longer term, the IMF could 'change the game' by building a multipolar system, he said."

That's right – no more US dollars as the world reserve currency. Carney hinted at a digital replacement perhaps to be administered by the IMF. That's the same IMF – the unelected IMF – which has dispensed its special brand of quack medicine on hapless developing countries around the world for the past seven decades. Got a financial problem? Call the IMF. Their debt-induced-asset-stripping-economy-destroying-indentured-servitude-shock-doctrine-neoliberalism-on-steroids-solution will get you back on your feet in no time. Or maybe it won't. Ask any African or Latin American country, or perhaps ask the people of Ukraine, officially the second poorest country in Europe after Moldova, how it's worked out for them since 2014?[22]

The IMF are the people Mark Carnage believes should be entrusted with administering the world's new monetary system.

In an interview with CNBC the same day Carney said,[23] "Now the issue is you don't just jump to something new overnight. And what we want in a multipolar world, I think we would agree we have a European engine, we've got the Chinese engine, we've got the US engine of this economy, a multipolar world, *you need a multipolar currency.* The question is how do you get there? And I laid out ideas of how you would get there." (Emphasis added.)

Just like that, the first shot across the bow of the dollar's reserve status was fired. Shrewd observers have been predicting this for years, but coming from the mouth of one of the world's most respected central bankers it would have come as a huge surprise to those who believe 'king' dollar's status could not be challenged. Most people can conceive of no other monetary system than the one where the US dollar reigns supreme but we can be certain that a new system is coming. After all, Mark Carney is telling us and one rather gets the feeling that he is privy to more information than the rest of us. Like the rollout of net-zero carbon, he believes it is urgent. He closed his speech with the words: [24]

"Given the experience of the past five years, I will close by adding urgency to Ben Bernanke's challenge. Let's end the malign neglect of the International Monetary Financial System and build a system worthy of the diverse, multipolar global economy that is emerging."

There will be millions of words written about the demise of the US dollar as the world's reserve currency in the not-too-distant future. For now, let us make two brief points. First, the US dollar's collapse is a mathematical certainty and it will be a cataclysmic event in economic history. Second, if one wishes to topple capitalism, it is a prerequisite that the US dollar be taken down as well. The US dollar simply has to go.

Mark Carney is not content dictating that businesses failing to adhere to the green rules handed down by Michael Bloomberg are to be 'punished'. This former Goldman Sachs banker has also declared that the post-World War ll financial system based around the US dollar needs to be replaced. In the past, when prominent figures threatened to disrupt the dollar's hegemony, they met with a sticky end. Saddam Hussein began to sell oil in euros instead of US dollars and quickly found himself at the end of a rope after his country was invaded. Colonel Gadaffi attempted to set up a gold-backed currency to free Africa from western monetary influence and his country was destroyed before he was brutally sodomised with a bayonet as Hillary Clinton cackled with glee. Mark Carney on the other hand? He shows up at a symposium in Wyoming as a Canadian citizen working for the Bank of England and announces that the dollar needs to make way for a new monetary system.

Who is this guy?

If Mark Carney's warnings and threats of bankruptcy can be considered the 'stick', he has also offered a carrot to businesses. Speaking at the Green Horizon Summit on 9 November 2020 he said[24i] "Achieving the net zero will require a whole economy transition, involving every company, bank, insurer and investor, and creating the greatest commercial opportunity of our time."

Mark Carney's audience at the Green Horizon Summit was not a bunch of tree huggers (with the possible exception of Prince Charles, who likes to talk to them). The summit was hosted by the City of London Corporation and there aren't many words in the English language that spell 'high finance and power' better than 'City of London Corporation'. The event was sponsored by the WEF and attended by business leaders from 100 of the world's biggest corporations. This was a business conference, not a meeting to discuss environmental issues.

Here was Carney telling a bunch of corporate leaders about 'the greatest commercial opportunity of our time.' These people exist to make money and language like this will have had the assembled bankers rubbing their hands. Think how much bankers and corporations have already skimmed off the top of the economy via quantitative easing and then imagine an even greater commercial opportunity.

By the way, does this sound like the death of capitalism to you? When Christina Figueres says capitalism is to be replaced, she means for us peasants, not their banker friends. There are certain people who intend to remain wealthy in perpetuity and add to their wealth. And who do you think Carney and Figueres really work for?

For all their speculative casino activities, banks do play an important role in the real economy. They finance regular businesses and they have been told quite specifically by Carney that 'every company, bank, insurer and investor' needs to be on board with the net zero agenda. Bankers have a simple choice reminiscent of George W Bush's declaration that you are either 'with us or with the terrorists'. Either you are with us regards net zero where bountiful riches await, or you are against us and can look forward to bankruptcy. How do think a banker will act in this spot?

The upshot is that banks will stop lending to anyone who doesn't sing from this carbon hymn sheet. The wealth won't be trickling down, but the net zero policies will. That is if the bankers know what's good for them, and I think they do.

Mark Carney wears yet another hat (in fact he has several others) that is worth a quick mention; this one in the private sector. In August 2020, he joined the Canadian firm Brookfield Asset Management Inc, a firm with over $600 billion under management, his role listed on the company website as Vice Chair and Head of Transition Investing.[24ii]

So it's back to making money for the 13 year Goldman Sachs veteran. "In this role, he is focused on the development of products for investors that will combine positive social and environmental outcomes with strong risk-adjusted returns," says the company blurb.

There is little doubt that Carney will earn his new company 'strong risk-adjusted returns' trading stocks in the Environmental, Social and Governance (ESG) sector. Given all his other roles and connections, the only surprise is that his new company's share price did not double on the news of his appointment. (That took a year.)

Can you conceive of a person more riddled with conflicts of interests for a job trading stocks in the ESG arena? Mark Carney knows everything about the sector, right down to which companies are ordained to succeed and which are earmarked for failure. Surely every single trade executed by one of Carney's staff has to be a candidate for insider trading? He is the Climate Envoy for the UN and self-appointed sheriff who dictates to businesses who will thrive or go under. He established the TFCD, the body that transmits global rules to obedient governments to implement, a body that is run by his buddy, Michael Bloomberg, while Carney simultaneously sits on the board of Bloomberg philanthropies. This man knows practically everything that is going on in the sector and he knows ahead of time what changes are afoot. Unless we are supposed to close our eyes and pretend that employees of the giant vampire squid have their fangs and tentacles figuratively removed on leaving the company, this is a serious ethical issue.

Not to worry, I'm sure his company compliance officer and those hard working regulators will be on top of everything.

Mark Carney left his Bank of England job on 16 March 2020, five days after the WHO declared a COVID-19 pandemic. Lockdowns across the world guaranteed a huge recession would follow, with enormous declines in GDP and rising debts that were already astronomical and which cannot possibly be repaid. We are in the early stages of the greatest ever depression.

Andrew Bailey replaced Mark Carney as head of the Bank of England and inherited an unholy mess, with interest rates at practically zero and no scope for cuts. But, let us not feel sorry for him and whatever you do, do not be tempted to believe Mark Carney is blameless.

1.4

They're Coming for Your Pension

Businesses are one thing. What about individuals? Will this affect us as well?

Read the following quote and decide for yourself. As we saw with the frank admission from Christina Figueres, once in a while politicians and people of influence make revealing statements that get to the heart of the matter. Here is another example. In December 2017, at the twenty-third session of the Conference of the Parties (COP 23) in Bonn, the COP 23 President and Prime Minister of Fiji, Frank Bainimarama said,[25] " ...after all, when we talk about tapping into the vast amounts of institutional capital for climate solutions we are largely talking about the retirement savings of ordinary hard-working citizens and *we need to honor the expectation of being good stewards with the money...*" (Emphasis added.)

Say what! The Prime Minister of Fiji is openly talking about spending our pension funds. I don't know about you but he didn't mention anything to me about being a good steward with my money.

That is people's life savings he is talking about. By this stage, most people are probably resigned to some kind of 'carbon' tax but having their pensions spent by politicians and UN bureaucrats too? Pensions should be sacrosanct. The prospect of a comfortable retirement is what keeps people going for 40 years on the hamster wheel. That is why pensions attract tax benefits and cannot be touched until retirement age. Now the climateers are saying even your nest egg is up for grabs. They haven't revealed their precise plans yet but they are talking about it in public which is enough reason to be alarmed. Public statements imply prior discussion in private and that means some kind of plan is underway.

Just occasionally, these comments and suggestions slip out that let us in on their line of thinking. This is a sensitive issue and we would not expect this message blared out from the rooftops. To the incredulous or those people who think everything is a 'conspiracy theory' please read that comment again and ask yourself how else it can be interpreted.

Treating other people's money as if it belonged to them is so typical of politicians. He speaks of being a 'good steward' with the money but he still plans on spending it. When these people speak of unlocking $100 trillion, bear in mind that in 2019, the GDP of the entire world was only $86.5 trillion.[26] It takes everybody on the planet more than an entire year to earn $100 trillion.

To those who think perhaps the quote is taken out of context, here are some other quotes from Frank Bainimarama's short speech of four key points:[25]

"Second – we must recognize the essential role of private finance and bring together both public and private climate finance. There are trillions of dollars sitting with private investment institutions, pension funds, investment banks and insurance companies. And they are all looking for opportunities to earn a return on these funds. *We must unlock that finance for our common cause.*" (Emphasis added.)

and

"Fourth – governments and the private sector must work together like never before with a sense of urgency and caution to unlock the trillions of dollars of private sector capital that can be mobilised to solve this challenge."

There can be no misunderstanding of this message. The Fijian PM is quite clear. They want to 'unlock' trillions of pension money to spend on climate change initiatives that may or may not work. Urgently. Bear in mind that the people who earned that money with decades of real labour are not even allowed to 'unlock' it themselves until retirement age.

How much does he believe it is all going to cost? Do they need 'trillions' to do whatever it is they are going to do? Will spending trillions of citizens' hard-earned pension money lower the world temperature? (If you have a brain stem you already know the answer to that one.)

None of this makes sense, added to which, nobody asked us if we wanted to contribute our life savings for the cause. Just for the record, I do not. I do consent to anyone 'unlocking' my pension, be it the unsolicited cold caller on my home phone, the esteemed Prime Minister of Fiji or the smiling duo of Mark Carney and Christine Lagarde. It is galling enough that my pension fund is invested by, administered by and 'guarded' in a custodian bank by people I have good reason not to trust, and that these institutions are regulated by people I know firsthand to be incompetent.

Further, in whose hands will all the unlocked money end up? Companies that do not conform to Michael Bloomberg's disclosure rules will be on the naughty step or at risk of bankruptcy Carnage-style. So it is a fair bet that the money will end up in the hands of 'well behaved' companies, especially those offering green 'solutions'. These are often the same companies who have been lobbying for climate change action and who partner with the WEF. It's a big club alright.

1.5

Corporate Green Muscle

It is the mistaken belief of many an environmentalist that the green movement is fighting a David v Goliath battle against an all-powerful fossil fuel industry with the greens playing the role of puny underdog. The reality is the exact opposite. The green movement is far better represented and funded than the fossil fuel lobby and it's not even close. Not only does the green movement have central banks, the UN, politicians and the media fighting their corner, but corporate investment also comes from all angles including a few places people might not expect.

For instance, did you know that the Paris climate summit of 2015 was partly funded by EDF, Renault, BNP Paribas and Air France?[27]

Most people would never have heard of the 'Creo Syndicate', a private members' club where ultra-wealthy investors meet to discuss ways to profit from green energy investments. Members come from 200 or so of the world's wealthiest families, the type whose influence transcends party politics. The group has over $800 billion under management. As Bloomberg reported on the Creo syndicate:[28]

"Creo acts a little like an investment bank, vetting about 300 deals per year, connecting investors with possible partners and conducting research on technologies. Members have invested in everything from batteries and hydrogen fuel to regenerative farmland and greener product packaging. Portfolios include still unproven technologies such as methods for carbon capture and true long shots like fusion reactors."

The group's founder, Régine Clément, is completely honest about its purpose:

"This is not philanthropy, this is investment."

Nor is there any requirement for members to divest from fossil fuel companies. These people are in it for the money and they know there is the political will to force these green projects to succeed, often in the form of government subsidies.

In the 2000s, Goldman Sachs was instrumental in devising dodgy carbon trading schemes along with Al Gore who made off like a bandit and is far from done milking the green cash cow.[29]

The green movement is not the underdog in this fight. In all truth, there isn't even a fight taking place, just the appearance of one. What should be even more alarming to a naïve environmentalist willing to learn some uncomfortable truths is that it is often the *very same corporations* the green movement so despises that are funding green initiatives.

Returning to the Bloomberg report on the Creo syndicate:[28]

"Part of building trust with wealthy families is keeping their secrets... the group's ranks include other well-known billionaires whose names Creo won't disclose... Some members in Europe have been rich for hundreds of years. Families 'are naturally inclined to think long term' she [Régine Clément] says."

One can probably deduce the identities of some of those members from that information.

Some of the richest family dynasties on both sides of the Atlantic are promoting the green movement. Isn't it interesting that the Rothschild[30] and Rockefeller[31] families, who have profited so handsomely from oil, gas and general planetary destruction, are in bed with the green agenda? You can't get more 'big oil' than Rockefeller and the Rockefeller Family Fund has even funded Alexandria Ocasio-Cortez's Green New Deal proposals.[32]

The banks, oil companies, hedge funds and the richest families in the world are all over the green movement. It's David and Goliath all right but the odds-on favourite in the 'fight' is the marauding juggernaut painted bright green. This is not about environmentalism. This is about money and power. Billionaires are going to make even more money and much of it will come from public funds. By 'public funds' we are not only talking about tax breaks and subsidies. Our pension funds will gradually be diverted into these green investments, but long after the likes of the Creo Syndicate have snaffled the early pickings.

Quoting again from Cory Morningstar:[33]

"On 8 November 2018, the article 'The Climate Finance Partnership: Mobilizing Institutional Capital to Address the Climate Opportunity' discloses where the money will come from for the 'fourth industrial revolution', sold to the public under the guise of sustainability:

So, what can be done? Whether you choose to look through the lens of unprecedented challenge or unprecedented opportunity, **there is violent agreement that institutional capital needs to be 'unlocked'** (a favourite word on the climate conference circuit) and **mobilised quickly and at scale**." (Emphasis in original.)

There might be 'violent agreement' within their echo chamber but not from me, or the majority of savers across the world one imagines. She adds:

"... the very foundations which have financed the climate 'movement' over the past decade are the same foundations now partnered with the Climate Finance Partnership looking to unlock 100 trillion dollars from pension funds.

To be clear: The money for multi-billion-dollar corporations – to create privatised services and industries, under the guise of environmental protection, is going to be paid for by the public – but the public will not own them. For the corporate sector, it's no risk – all profit. Anything that fails – the public is on the hook.

By utilizing the non-profit industrial complex, the world's most powerful oligarchs need not force their will onto society. Rather, akin to what Aldous Huxley prophesized in his fictional novel *Brave New World*, we have been manipulated and engineered to demand the very 'solutions' that will further empower those that destroy us."

This is the magic of using Greta Thunberg as a figurehead for the movement. By persuading people to get behind Greta, they will not only accept a diabolical economic plan that impoverishes them while further enriching billionaires, but they will also embrace it. They will actively demand it.

1.6

Human Shields

In this connection, it does not help that the climate change movement has taken on almost religious fervour. Greta is like a spiritual leader to some people and media outlets have even called her 'The Joan of Arc of climate change' without irony.[34]

Using a child as the figurehead of a campaign has another benefit; it is bad form to criticise a child, especially one with Aspergers. This is something Greta's promoters take maximum advantage of.

Not that 99.99% of people would even want to criticise Greta. Child activists are like vegan dogs – you know who is making the decisions. Many see her as an unfortunate puppet who is being used to push an agenda, this author included. Dare to point this out though, and you are a killjoy, or worse, you do not care about the planet!

A lot of people self-censor if they have anything negative to say about the whole Greta Thunberg phenomenon. Criticism of the media circus, school skipping, or Greta's relentless schedule – no matter how valid – is met with harsh rebukes and turned into a direct criticism of Greta Thunberg herself. Wicked motives are attributed to perceived attacks on Saint Greta. People have been accused of all sorts for daring to express an opinion that falls short of 100% unwavering support: misogynist, climate change denier, disablist, Trump supporter (of course) and even 'white nationalist' (go figure). Criticism of the circus is met with the standard refrain:

"Why do you hate Greta?" (Imagine a whiny, scornful voice saying it while peering over the top of *The Guardian* newspaper.) The mainstream media has been relentless in pushing this line. Here is a selection of headlines:

"Why are powerful men so scared of Greta Thunberg?" (*Independent*)[35]

"Why is Greta Thunberg so triggering for certain men?" (*Irish Times*)[36]

"Greta Thunberg's defiance upsets the patriarchy – and it's wonderful." (*Guardian*)[37]

"Why middle-aged men hate Greta Thunberg." (*Financial Times*)[38]

"Greta Thunberg's enemies are right to be scared." (*Guardian*)[39]

It's almost funny.

As to Greta's 'enemies', who are these imaginary people? Norwegians? Vikings? To all these moronic headline writers penning their straw man headlines and asking why evil white men hate Greta so much, newsflash – they don't. Perhaps some of them are just able to see through the scam? This is probably what they find so upsetting at Guardian HQ and the other organs whose job it is to manufacture hate out of thin air. As an aside, it is hard to debate rationally with people who attribute all the opinions they do not like to 'hate' of some description. They'll accuse people of either hating women, hating kids or hating the environment. Perhaps hate is the emotion these people are the most intimately connected with, so it becomes the first one they attribute to others? No rational people hate Greta Thunberg. They might hate the manipulation of her but headline writers tend not to let that get in the way of their politically correct virtue signalling.

While this propaganda is so bad that it only tends to work on the brain dead, it does succeed in getting people to self-censor. Criticise any aspect of the Greta Thunberg circus or the Green New Deal, and you are a child-hating misanthrope who doesn't care if the Amazon burns down. Most people who can see through it do not want the 'misogynist', 'disablist' or 'hater' label so they keep schtum. Better to keep quiet and accept the whole package; lock, stock and barrel. Greta, the Green New Deal and, by extension, the tearing down of capitalism, although most people are not aware of the last part.

This kills intelligent debate, nipping it in the bud. One rather gets the impression that this is the whole point. It is as if Greta is the human shield who confers protection from any criticism of the Green New Deal and stops people from looking behind the curtain to see who is pulling her strings.

Mainstream media has failed us. The media has an obligation to be objective, to analyse and to criticise where criticism is due. It is their job. For the most part, they have acted as cheerleaders and cynically positioned themselves as a self-appointed child protection unit for vulnerable Aspergic teenagers.

It is downright hypocrisy. If they cared about child welfare they would be calling this charade out for what it is, perhaps questioning the wisdom of feeding an already obsessive child with visions of apocalyptic, planetary doom. They would also be questioning whether any of these predictions stand up to scientific scrutiny. In other words, doing actual journalism, ie their jobs. Don't hold your breath.

When you look across the range of mainstream media coverage, you cannot fail to notice how one-sided it all is. Major TV channels give Greta Thunberg and the school strike protests total support because, almost to a man, they agree it is all for a good cause. The message is uniform: 'climate change is terrible. Manmade CO_2 is causing it. We need to slow/stop/reverse it.' We have been inundated with this message for decades now.

If one never criticises the premise of the protest, one will end up agreeing that the protest itself is a good thing. This is the risk if the media all speak in one voice without an alternative viewpoint. (One of the only notable exceptions is Sky News, particularly Sky News Australia.)

This bias has been codified by the BBC. On 6 September 2018, guidance was issued to BBC journalists on how to report climate change issues.[40] From that date on, it was not even necessary when covering a climate change story to invite someone who would argue the other side of the debate. The infamous four-page 'crib sheet' included the following:

"Be aware of 'false balance'. As climate change is accepted as happening, you do not need a 'denier' to balance the debate. Although there are those who disagree with the IPCC's position, very few of them now go so far as to deny that climate change is happening. To achieve impartiality, you do not need to include outright deniers of climate change in BBC coverage, in the same way, you would not have someone denying that Manchester United won 2-0 last Saturday. The referee has spoken."

Bear in mind that the BBC is a state broadcaster funded by a licence fee levied on viewers. People who do not pay the licence fee can be sent to prison and often they are. In light of this, the BBC is obliged to be impartial and this obligation is written into their charter.

Most people would be hard-pressed to name a single person who claims the climate does not change. The fact we had ice ages is evidence of climate change. There are millions of people, however, who agree that climate change is happening and who also agree that this is due to greenhouse gases but that it is not *all* caused by man. There is a whole spectrum of opinion in between the two poles. Some say all the change is caused by man, some agree that climate is changing but none of it is caused by man and some say a proportion of it is caused by man, which could vary from between 0-100%. Claiming the issue is as settled as Manchester United winning 2-0 and that 'the referee has spoken', is a gross over simplification. The debate is not settled by a long way and never will be while zealots and critical thinkers co-exist.

The BBC presents this as though they are avoiding 'false balance' as if 'balance' was something they held to be sacred. As to the corny punch line 'the referee has spoken' it appears that the BBC not only speaks to its viewers like ten-year-olds but their own staff as well.

The climate is changing. Hard-core deniers are wrong to say this is not true. Therefore, the BBC argues, producers do not need to invite a climate denier to provide balance. On the face of it, this would be OK if this only extended to outright deniers, who probably number fewer than one in 10,000 people. The problem is that this is extended to sceptics of all kinds, which means all the reasonable sceptics with valid questions are excluded too.

The BBC crib sheet states the following:[40]

"What's the BBC's position?

Man-made climate change exists. If the science proves it we should report it. The BBC accepts that the best science on the issue is the IPCC's position." (The Intergovernmental Panel on Climate Change.)

The statement that 'The BBC accepts that the best science on the issue is the IPCC's position' will draw howls of derision and outright laughter from those who have kept an eye on these matters over the years. Many former IPCC scientists also beg to differ in the strongest possible terms. Contrary to conventional 'wisdom', the IPCC is not a disinterested observer that studiously reports the facts without bias. The IPCC is a body of the UN and acts more like a political body than a scientific one. It was set up in 1988, with the express goal of proving that man was to blame for climate change. Small wonder that this is the exact thing it has 'concluded'.

Many IPCC scientists do not even support the view that climate change is manmade. You read that right. Take Dr John Christy for example. He is the Professor of Atmospheric Science and Director of the Earth System Science Center at the University of Alabama. He is also the winner of a NASA Medal for Exceptional Scientific Achievement for his work on the global temperature data set from satellites. Writing in the Wall Street Journal in 2007 this distinguished former member of the IPCC said, "I'm sure the majority (but not all) of my IPCC colleagues cringe when I say this, but I see neither the developing catastrophe nor the smoking gun proving that human activity is to blame for most of the warming we see."[41]

Oh dear.

How about another former IPCC member, Dr Willem de Lange, Senior Lecturer in Earth Sciences at the University of Waikato, who said, [42] "In 1996 the IPCC listed me as one of approximately 3,000 'scientists' who agreed that there was a discernible human influence on climate. I didn't. There is no evidence to support the hypothesis that runaway catastrophic climate change is due to human activities."

Ouch.

Dozens of scientists have left the IPCC in disgust. Whistleblowers have complained that all too often scientists are not chosen not for their competence but their ideology. They have laid waste to the notion that the IPCC recruits the world's best scientists to look into climate change. Many are extremely young and not even properly qualified, with lead authors in their mid-20s having not even started their PhDs. It has been proven that IPCC reports contain hundreds of references to studies that are not peer-reviewed and even to newspaper and magazine articles. According to some of its former scientists, the IPCC is not an impartial body and a purveyor of junk science.[43]

Here's a quote from another former IPCC expert reviewer, Dr Richard Courtney:[44] "To date, no convincing evidence for AGW (anthropogenic global warming) has been discovered. And recent global climate behavior is not consistent with AGW model predictions."

It is very difficult to take that quote out of context. The same sentiment goes for the 46 statements made by former IPCC members at the end of this chapter. Next time you see some journalist bowing and scraping at the feet of the mighty IPCC, have a read of these 46 comments and laugh.

Of those comments, perhaps one of the most valid criticisms was made by Danish physicist Dr Eigil Friis-Christensend:

"The IPCC refused to consider the sun's effect on the Earth's climate as a topic worthy of investigation. The IPCC conceived its task only as investigating potential human causes of climate change."

The IPCC has repeatedly failed to address the effects that the sun is having on the Earth's climate. The grievance expressed that the IPCC only looked at human causes ought not to surprise us; it is the whole reason the IPCC was established.

But ignoring the sun is a catastrophic omission for any serious analysis of climate. After all which is the more plausible scenario?

1 - that the sun's output is constant and man controls the temperature on earth with his greenhouse gas emissions like a thermostat?

2 - that climate change is largely influenced by the sun, an enormous fiery mass that could fit 1.3 million Earths inside it and whose output is not constant, something you can tell just by looking at it.

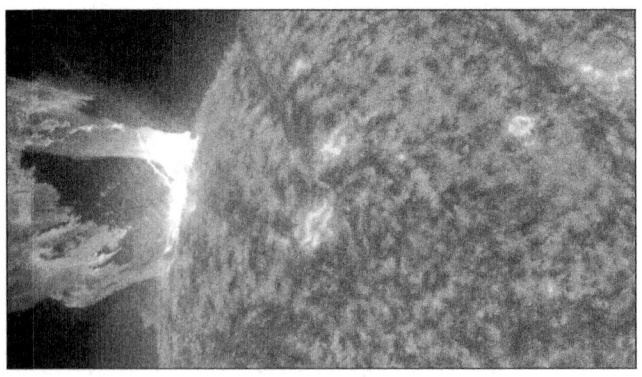

A possible explanation for climate change?

Indeed, the IPCC's scientists are world-class experts at ignoring the effects of the sun. It is as if they can pretend it does not exist. It's getting to the point that the IPCC could be classed as a 'sun denier'.

Using the IPCC as the 'best science' is like using the fans' forum of Liverpool Football Club website as 'best evidence' as to who the greatest football team is. By putting the IPCC on a pedestal, the BBC is just making one giant appeal to authority and asking us to trust them.

By the way, if the science is 'settled', as so many people like to claim, why are climate predictions so useless? Climate change models have never proven to work with any accuracy yet still we are bombarded with doom and gloom scenarios, often with deadlines attached. Twelve years to save the world, no 8 years, no 3 years, famine by the year 2000, millions to die of starvation, sea level rises to flood country X in 10 years and so on. Al Gore, Prince Charles, Greta Thunberg; the faces change but it's the same old story and the scary scenarios never materialise.

This is more a criticism of the media than the IPCC. The IPCC knows full well you cannot predict climate states into the future. Did you know that in 2001 the IPCC said in its Third Assessment Report that:[45]

"The climate system is a coupled non-linear chaotic system, and therefore the long-term prediction of future climate states is not possible."?

There you have it – from the IPCC in its own words. That is the body the BBC relies on for its 'best science', right there. 'Long-term prediction of future climate states is not possible.' Well, this is awkward.

Climate scientists like to take a model, back-test it against real-world data and then alter a few parameters such that the output coincides with what happened in the real world. This is just statistical chicanery. Fudging the models with hindsight to make them work might fool a few people for a while but it is an almighty leap of faith to assume the model will work in the future in a chaotic system.

As the saying goes, *an economist is someone who will tell you tomorrow why the prediction he made yesterday did not come true today.* Are climate scientists the new economists? Perhaps they are even worse. One imagines that if they tried their hand on Wall Street they wouldn't last a year, yet they are lavishly funded with taxpayer and foundation money to spend their careers producing models that politicians and technocrats convince the media to scare us with.

So far we have had about one degree of warming in the last century which, in the scheme of things, is just noise in the data. These climate models will not be proven right or wrong for decades or even centuries. Until then, the models are just theories. It is far from the truth that climate science is 'settled'. It can't be, at least not yet, nor for a long time.

If long-term predictions are not possible, it renders all this wailing and gnashing about "12 years", "8 years" and "3 years" as absurd fantasies. They are.

Have you noticed that when predictions fail to materialise, the alarmists just make even more extreme predictions? The media dutifully reports these apocalyptic predictions and presents them to viewers as 'the science'. One last quote from the crib sheet shows what a mockery the whole Greta Thunberg saga makes of the BBC:[40]

"As with all topics, we must make clear to the audience which organisation the speaker represents, potentially how that group is funded and whether they are speaking with authority from a scientific perspective – in short, making their affiliations and previously expressed opinions clear."

Did the BBC do *any* of this with Greta Thunberg?

You would think that the average person already knows whether or not Greta is 'speaking with authority from a scientific perspective' but judging by the number of grown adults who sit in awe of Greta and her message one cannot be certain.

Let us spell it out just in case the BBC omitted to tell us; Greta Thunberg is scientifically illiterate. This is not meant as an insult but as a statement of fact. *Clearly*, Greta is not 'speaking with authority from a scientific perspective'. She doesn't go to school. She claims to be able to see CO_2, which is invisible. Strictly speaking, her mother made this claim on Greta's behalf in a book, something Greta backtracked on as 'a metaphor taken out of context'. However, the claim was quite specific in that Greta could see CO_2 with the 'naked eye', so she doesn't know what a metaphor is either. World leaders and policymakers should not be taking advice from this girl.

As to keeping their audience appraised of who is funding the Greta Thunberg movement, what has the BBC done to explain the machinations behind her 'meteoric' rise? There is a veritable industry of companies and NGOs pushing her agenda and it is a vastly better-funded lobby group than the fossil fuel lobby. Does the average BBC viewer know this? One would guess not, but if they are aware it certainly isn't because the BBC informed them.

Why is it that unpaid researchers are having to do this work in their spare time? What are people paying their licence fee for? Why do people believe a word the BBC says? The BBC is possibly the most famous broadcaster in the entire world yet it neither follows the commitment to impartiality enshrined in its charter nor its own (incredibly biased) internal guidance.

This is the extinction of journalism.

1.7

Playing the Ball Not the Mann

If you think the accusation of bias against the BBC is unfair, there is a recent example which even the most passionate environmentalist would struggle to deny.

Consider the example of Michael Mann's 'hockey stick' graph which took the world of climate science by storm in 1998. It was given the name because the graph representing northern hemisphere temperatures elevates steeply at the end of the thousand years, thus resembling a hockey stick. The inference drawn by many is that human behaviour must have caused a temperature rise when fossil fuel burning caused so much CO_2 to be released into the atmosphere.

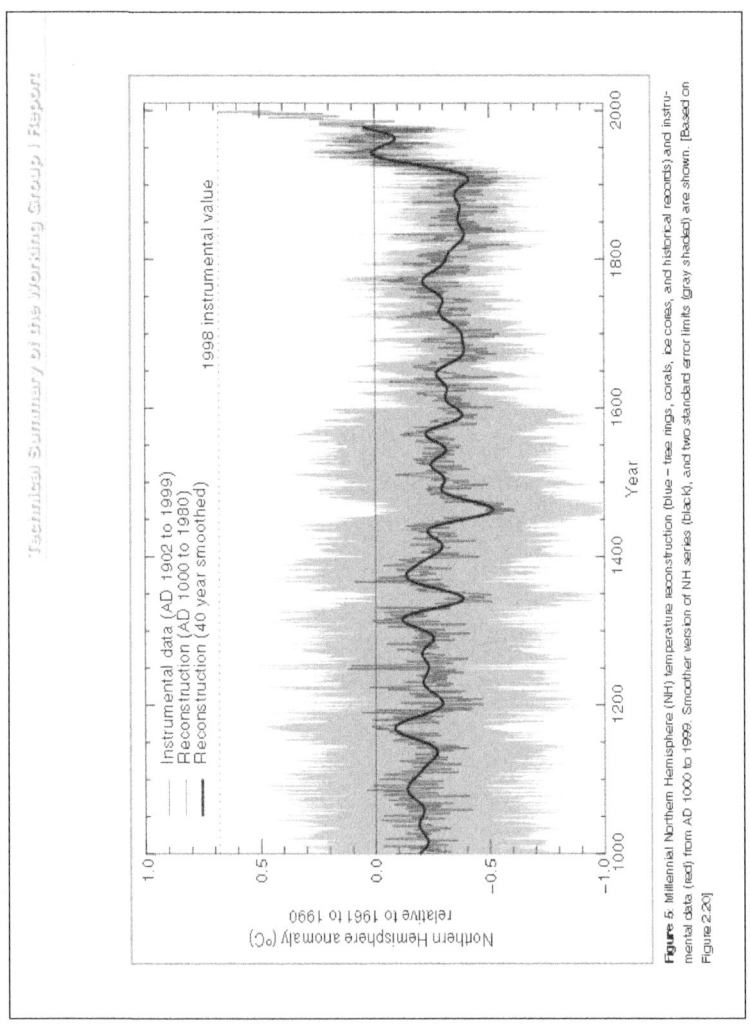

Michael Mann's hockey stick graph as presented in the IPCC Third Assessment Report (TAR) Working Group 1 (WG1): The Scientific Basis.

The hockey stick graph was featured in Al Gore's 2006 film 'An Inconvenient Truth', which won an Oscar for Best Documentary but unfortunately contained nine scientific errors according to a British judge in 2007 (other observers claim there are as many as 35 scientific errors).[46]

With there being no thermometers centuries ago, proxy readings were estimated from tree rings, corals and ice cores. The graph shows an extended steady period before temperature increases 'violently' by one whole degree in the last century. And voila – we have a scary graph to alarm the public and influence policymakers.

It certainly did both of these things. Mann's graph was an essential part of the IPCC's 2001 Third Assessment Report referenced above. It was featured prominently in the *Summary for Policy Makers* section and as the name suggests, policymakers use this information to base their policies on.

But there was a problem. It is widely accepted that there was a medieval warmer period in the northern hemisphere from about AD 1000 to AD 1300 where temperatures were higher than they are today. What happened to it?

One curious and very talented Canadian mathematician set out to replicate Mann's graph and find out the answer to this question. In 2003, Steve McIntyre requested Mann to produce his data sets and methodologies which ought to have been a simple matter for honest people conducting actual science. Instead, McIntyre faced non-compliance and hostility from Mann. McIntyre started a blog called Climate Audit to document his efforts in reconstructing Mann's graph.[47]

Then 'Climategate' happened in 2009. Following a computer hack, thousands of private emails between scientists at the Climate Research Unit at the University of East Anglia were released shortly before the Copenhagen Summit on climate change. It was an international scandal and rightly so. It should have been an even bigger scandal but over time it has been downplayed and its importance diminished. Do not be fooled by lame attempts to whitewash the seriousness of Climategate, or be tempted to agree with handpicked reviewers who concluded that emails were taken out of context or cherry-picked. Read them for yourself.[48] Some of these emails included information that showed how the hockey stick graph was constructed. Amazingly, these emails showed that inconvenient data had been removed to create the hockey stick.

Enter Dr Tim Ball and eight years of litigation, which only ended on 22 August 2019. In February 2011, climate scientist, Tim Ball, was giving an interview in which he quipped, "Michael Mann should be in the State Pen, not Penn State." Ball was referring to Mann's conduct concerning the hockey stick graph. In March 2011, Michael Mann, the Penn State University Professor, sued Tim Ball for libel in the Supreme Court of British Columbia in Vancouver based on this quote.

Finally, on 22 August 2019 – more than eight years later – the British Columbia court dismissed Mann's claim with prejudice and awarded court costs to Ball.[49] 'With prejudice' means that Mann has no right to reinstitute the case.

Those who support Michael Mann will point out he only lost a libel case. However, the details of the case and what followed are damning, not only for Michael Mann personally but also the whole climate change movement, in particular the UN body acting as the guardian of climate change policy, the IPCC.

Michael Mann lost the case because he ultimately refused to provide all his data sets and methodologies as part of the discovery process, in particular his regression analysis. Yet he was the one bringing the case! As the plaintiff, he had to have known that the first thing Ball's defence team was going to do was request the very data that Mann had so zealously been resisting making public.

Yet that is what happened. The court gave Mann every opportunity, which helps explain the eight-year delay. In 2017, the case was supposed to go to an initial trial but this did not happen as Mann continued his cynical stalling tactics. Presumably, the judge lost patience with Mann, dismissing his claim for libel with prejudice. One imagines Michael Mann would not even want to appeal because he would only be asked to provide his data sets again, something he is never going to do.

Perhaps if he did, his hockey stick graph would not be quite so hockey stick-shaped and might resemble Tim Ball's version more closely.

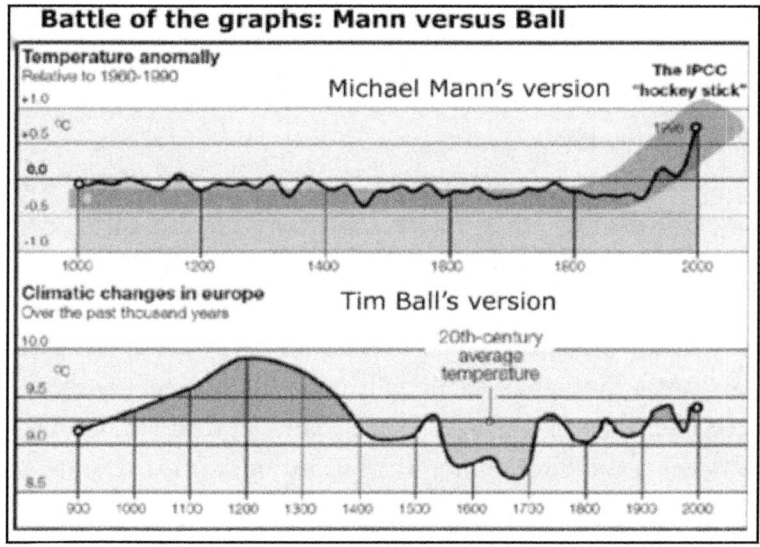

Regardless of one's opinion on the validity of the hockey stick graph, the fact is that Mann refused to give up the data showing how he produced it, proving that both he and his graph are unreliable. Mann evaded the truthful path and so does anyone else who goes along with the story his graph depicts, which of course includes the IPCC.

After his legal defeat, Mann took to Twitter on 25 August 2019, to put an interesting spin on the result:[50]

"BC Court never made any finding that I failed to produce any data. Anyone claiming otherwise is lying or facilitating a lie. Ball got the case dismissed based on: 1. Complaints about his age and poor health 2. Arguing his attacks on me were impotent anyway."

Mann's first sentence takes the biscuit after more than a decade of stonewalling. The entire debacle only came about because Mann refused to produce his data! The court judgement blamed Mann for delays but just because it did not specifically mention his failing 'to produce any data', Mann leapt upon this and somehow tried to use it in his favour.

The judge might not have specifically found that Mann 'failed to produce any data'. Mann blatantly did fail to produce vital data though – for years and years on end – and everyone knows it, including him.

In relation to Ball's ill health, Mann wrote another tweet on the same day that read:[50]

"What does it say about Tim Ball's honesty (and that of the right-wingers trumpeting the latest court development) that they conveniently failed to note that Ball sought to dismiss the case based on his alleged health problems."

Tim Ball suffered coronary heart failure in 2017, aged 79, already six years into the legal battle. He had undergone surgery for a quintuple heart bypass surgery a decade before. With his health issues and his opponent cynically stalling in court, it was absolutely fair to ask the judge to take into account his failing health. He should have been enjoying his retirement instead of being financially drained by the legal tactics of Michael Mann. No wonder the judge saw fit to draw a halt to proceedings in the interests of justice. Had Tim Ball died before the case was concluded it would have benefitted Michael Mann immensely.

Yet here was Mann asking what it says about Ball's honesty that he 'sought to dismiss the case based on his alleged health problems'. What does it say about Michael Mann that he refused to provide the data for years – both as a scientist and a man? What does it say about Michael Mann's integrity that he referred to Tim Ball's health problems as 'alleged'? A quintuple heart bypass! And what does it say about Mann's honesty that after all the legal shenanigans he indicated he wouldn't even pay the reported $700,000 costs?[51]

Note also Mann's logical fallacy of labelling Ball's supporters 'right wing', as if it had any relevance to anything, even if true. This smear tactic is the hallmark of one who does not act in good faith.

Now let us return to the BBC and take a look at some of their headlines regarding the hockey stick over the years:

"Climate legacy of 'hockey stick" – 16 August 2004[52]
"Backing for 'hockey stick' graph" – 23 June 2006[53]
"Climate 'hockey stick' is revived" – 1 September 2008[54]

All these articles were published after Steven McIntyre had cast serious doubt on the mysterious removal of the medieval warming period. While the articles do at least mention the fact there is *some* controversy around the subject, it is hardly the prominent theme and the headlines do not even hint at the raging disagreement or implications of scientific fraud.

Strangely, you will not be able to find a BBC headline from August 2019 reporting that Mann's libel case was dismissed let alone a mention of his refusal to supply data or what the implications of this are. Try searching for 'BBC Michael Mann loses case' either within the BBC website itself or using your search engine of choice. Perhaps it was just me – maybe my search engine is broken? Either that or the BBC is not following its impartiality guidelines.

The BBC did however interview Michael Mann on 23 January 2020 during its 5 Live Science radio programme.[55] After a fawning introduction and much praise from the BBC's presenter, the first 40 minutes contained plenty of general chatter but no mention of the case against Dr Tim Ball.

As an aside, during this section, Michael Mann made the evidence-free assertion that the recent Australian bush fires were a result of manmade global warming, an assertion that went unchallenged by the BBC. It wasn't global warming that made the 2019 Australian bush fires so bad, manmade or otherwise. Ask any Australian fireman what the problem was and you will get the same (probably quite blunt) answer; increased fuel loads. Any sane observer or person without an agenda will tell you the same thing.

Australia has always suffered from bush fires, but laws preventing the removal of fuel loads in national parks and preventing landowners from clearing local fire hazards have increased these fuel loads to the point that they result in raging infernos when lit. Guess who pressed for the introduction of these laws?

That would be the green lobby and the environmentalists. Now, some of those very same people have the nerve to blame manmade global warming for their own catastrophic and deadly stupidity. As the volunteerfirefighters.org website notes:[56]

"Every bushfire inquiry since the 1939 Stretton Royal Commission, has urged the use of systematic controlled burning, or 'prescribed burning', to burn off flammable ground cover in cooler months, so it does not fuel summer bushfires."

Indigenous populations have overseen controlled burns for tens of thousands of years, but obviously, the latte swilling, dilettante green lobby knows better than the people who have looked after the country for millennia.

Returning to the BBC interview with Michael Mann, in the 41st minute (of a 47 minute programme) the host said he had a question from a listener. This is what the BBC host said to Michael Mann, verbatim:[55]

"Let's ask you about something else now, that *one* of our listeners, er, knows about. *Again* I know nothing about this. Could you ask Dr Mann about his climate data case in British Colombia?" (Emphasis added.)

The emphasis on the words '*one*' and '*again*' is because the host himself put verbal stress on those words which sounded unnatural. Why he said 'again' is a mystery as the issue had not been mentioned up to that point. That the host of a BBC radio science programme went out of his way to profess ignorance on this matter was surprising, to say the least. Apart from the fact climate change is such a high profile issue he ought to have done his basic research on his guest. Why did he feel the need to deny knowledge of the case?

Michael Mann did not even mention Dr Tim Ball's case in his short answer. He spoke of a different libel case, one where he claimed victory, sounding unsteady and evasive as he spoke. The BBC host jumped in quickly to confirm it was 'only' a libel case and not a case about scientific work, feeling the need to make this distinction about a case he claimed he knew nothing about just one minute earlier.

My instincts on hearing this disgrace of an interview were to write a letter of complaint to the BBC but what's the use in wasting one's time just to be told that they think they are doing a fine job and in their own opinion they are unbiased? In any case, I haven't paid the licence fee for five years.

1.8

No Good Deed Goes Unpunished

There is a postscript to the Michael Mann story that beggars belief. A real scientist will always make his work available for independent scrutiny. A real scientist will actively applaud the man who can disprove his work, no matter how dearly he holds his findings personally. Of course, Michael Mann failed to do these things. Real scientists ought to regard Michael Mann as having disgraced the profession but that is a far cry from what transpired.

On 17 February 2018, the American Association for the Advancement of Science (AAAS) presented Michael Mann with its prestigious 'Public Engagement with Science' award, stating:[57]

"The honor recognizes Mann's 'tireless efforts to communicate the science of climate change to the media, public and policymakers.'"

Although the award was given before the libel trial had ended, Mann's conduct had already been exposed during the Climategate revelations and he was stonewalling Tim Ball's request for data, thus preventing discovery in the case by not revealing his methodology – the antithesis of good science. The AAAS, which is the biggest professional association of scientists in the world with 120,000 members, also said:

"In the past year, Mann has had 500 media interviews and appearances and directly reached public audiences via social media… He has also advised actor Leonardo DiCaprio, who spoke about climate change during a 2014 speech delivered to the United Nations."

Is being a whisperer to Leonardo DiCaprio something the AAAS holds in such high regard? Let us hope this is not indicative of their priorities. Surely even the scientists who believe in the most extreme climate change predictions must be embarrassed to call themselves a member of that association? And for a man who gave 500 interviews in a single year, Michael Mann sounded remarkably shifty and unconvincing.

It gets even worse. A glance at Michael Mann's website states:[58]

"Michael Mann, distinguished professor of atmospheric sciences at Penn State, has been elected to the National Academy of Sciences."

Handing out awards is one thing but being elected to the National Academy of Sciences (NAS) is something else. Perhaps apart from being awarded a Nobel Prize, being elected to the NAS is the crowning achievement for an American scientist. The NAS was established by an Act of Congress and its role is to provide advice to the government on scientific issues. Members of the NAS really can influence the USA's policy on science, for good or bad.

New members of the NAS are elected by existing members. Penn State news proudly informs us: "This year, the academy elected 120 members and 26 international members to its membership. Mann's election brings Penn State's representation to 16 members, and total membership in the academy to 2,403 active members and 501 international members."[59]

Mann's election to the NAS was in April 2020, after he had lost his libel case. As they say, no good deed goes unpunished.

It is a struggle to believe that Michael Mann's election did not run into strenuous objection from certain quarters of the NAS, namely those quarters where people still have integrity and believe in the scientific method or even that it is right and proper to pay costs awarded after losing a libel case. After all, Michael Mann completely discredited himself and disgraced the scientific profession. Given everything we know about the Michael Mann saga, it is also a struggle to believe that all is healthy in the world of climate science and that politics doesn't play a major role.

Awarding scientific honours to Michael Mann after his display in the libel case is like giving a Nobel Peace Prize to Barack Obama and Henry Kissinger. Oh, wait.

Indeed, it is like awarding an Oscar for Al Gore's error-strewn film 'An Inconvenient Truth' ('error' is being polite). Gore shared the 2007 Nobel Peace Prize with the IPCC and heard the announcement one day after a judge ruled his film was full of scientific baloney. Interestingly, the judge ruled that the film could still be shown to school children, so long as it came with a disclaimer saying it was not an impartial analysis. As if little kids are going to remember the disclaimer and not the film.

The awarding of prizes is a time-honoured tradition when forging a propaganda narrative. It is harder to criticise a corrupt police officer if he has 11 citations for bravery. Giving Jimmy Savile a knighthood certainly seemed to protect him from a proper criminal investigation. Helpful journalists are often showered with awards, only for it to transpire that they were being paid by the CIA all along. The issuing of awards helps to forge a narrative that shields people from valid criticism.

It seems that Michael Mann can do no wrong in the eyes of the scientific establishment. It also appears that a narrative is being manufactured that climate change is a real and present danger. Threats to this narrative, such as Michael Mann losing his case, are ignored by the media and, bizarrely, actually rewarded. Meanwhile, the real heroes like Steve McIntyre and Tim Ball, who told the truth and were completely vindicated, have had their names dragged through the mud and are labelled as pariahs and trouble makers with dubious motives.

Science has failed us. The BBC has failed us.

The reader might notice a lot of flak is heading in the BBC's direction but we could say similar things about most mainstream media outlets. Consider how the rest of the mainstream/corporate media presented the meteoric rise of Greta Thunberg. What words would you use to describe their coverage: 'enthusiastic', 'supportive', 'uncritical' perhaps?

It was presented as a spontaneous protest that just went viral as if it was a grassroots movement. We have seen that it was a most cynical and carefully prepared piece of manipulation involving billionaires and PR companies with the blessing of the UN and Central Bank heads. There is no way that entire offices of trained journalists can have failed to realise this. This is propaganda.

1.9

Ship Ahoy

Take the industry's reporting of Greta Thunberg's long haul journey in a specially crafted yacht for example. When Greta sailed across the Atlantic to attend the Climate Action Summit in New York, the yacht was hailed as a zero-carbon vessel. It was far from 'zero-carbon' however, being made from high-tech carbon fibre. The energy used in assembling the yacht from its raw materials, the production, moulding and construction would have been substantial. In addition, it was reported that the crew would fly back to Europe after the voyage, adding thousands of air miles to the tally. The whole project had a massive carbon footprint.

The yacht had an interesting back story itself, having been used by the racing team Gitana, founded by the French banker, Benjamin Rothschild. Gitana used to race the yacht with the name 'Edmond de Rothschild' plastered across its sail. It is hard to conceive of a more elitist name than Rothschild and it certainly puts the image of the regular school girl into perspective. One might have imagined Greta's handlers would be a little more circumspect.

There were also accusations of a chase boat. I do not recall seeing any aerial footage of the yacht making its way across the ocean, which seemed a bit strange, particularly as flying drones are so widely available today. One would have thought this would have made great footage for the media. Not so great if there was a chase boat in the picture though.

If there was a chase boat, Greta may have inadvertently given the game away with a picture she posted on social media from the Ark of Greenness itself. Can you see that shape in the top right corner (bright red in the original colour photo)? It is not definitive proof of a chase boat but it feeds suspicious minds.

Point out that this eco stunt is not as green as it is cracked up to be and you risk running the gauntlet of scornful indignation from her almost religiously fanatical supporters. "Why are you spitting in the faces of children who are trying to save the world!"

This is not to say there was a total absence of media scrutiny. Some questions were raised because they had to be. The yacht was not carbon-zero and the public are quite well educated on these matters, ironically because of the relentless green agenda. Yet one cannot escape the feeling that the media were acting as cheerleaders and not reporting the news dispassionately, which is their job.

This is not to say anyone told an outright lie. The man on TV did not look into the camera and say, "This is a grassroots campaign, all totally natural dude." Nobody said, "Greta is only funded by her parents and receives no support from NGOs linked to billionaires, on my life guv'nor." They don't need to lie to create and perpetuate a narrative. They just omitted to tell us the important details.

There might be some breaking of ranks after a while. The odd journalist might call into question the wisdom of piling this responsibility on a young kid with a disability or perhaps mention pushy parents and so forth. But, dissent will only be within the 'prescribed lanes' as it were. There can be no questioning the actual cause – that the world needs to lower its CO_2 output.

People might ask, "What does it matter? If we all emit less CO_2 and pollute less, the whole world will benefit, right? Even if the temperature doesn't fall the world will be cleaner."

This is exactly how we have been conditioned to think, but it is wrong. Being misled matters all by itself, even if no real-world consequences resulted from the deception. It is a form of dishonesty. Be certain that when influential people are trying to shape your thinking on a scale of this magnitude, it matters. And it just so happens that there are real-world consequences.

By now the majority of people will be a bit more cynical of the Greta Thunberg phenomenon than they were at first. But with propaganda, it doesn't matter. Just as the shocking but libellous front-page story is apologised for three months later at the bottom of page 14, so Greta Thunberg will become a historical footnote in the mists of time. Should Greta ever suffer psychological damage from being cynically used, her handlers and beneficiaries will be long gone. By that point, their mission will have been accomplished. Much already has.

It is what happens in the heat of the moment that counts. In 1991, the allied coalition went to war with Iraq, a war justified to the public largely on the basis that Iraqi soldiers had thrown babies from incubators onto the hospital floor in a Kuwaiti hospital and left them to die. It was a complete lie. The 15-year-old girl who gave this tearful testimony was Nayirah al-Sabah, the daughter of the Kuwaiti ambassador to the USA. The whole sordid scheme was invented by a public relations agency, Hill + Knowlton, who were paid millions by the Kuwaiti government to mount a campaign for strong action against Saddam Hussein. (Note that Amnesty International initially corroborated this fabrication. It is one of many times where Amnesty International has disgraced itself and it will not be the last.) The lie was exposed in due course but by then it was too late for the Iraqis.

It will be too late for the public by the time they realise the media conned them about Greta Thunberg. Much has already been accomplished. Take the UK for example. On 27 June 2019, a law was passed pledging to bring all UK greenhouse gas emissions to net-zero by 2050.[60] As part of this, the British government says it will ban the sale of all new petrol, diesel or hybrid cars in the UK from 2035.

This means replacing the entire vehicle supply with electric vehicles that use lithium batteries. Even ignoring the facts that lithium batteries are not environmentally friendly to produce, that they degrade over their lifetime (meaning fewer miles per charge) and that they function poorly in cold weather, there is absolutely no way the UK will be able to power 30 million or more electric cars on top of existing electricity demands. The national grid will not be able to cope with the strain that will be placed upon it especially when coupled with plans for mass electrical home heating. Either the UK will experience power cuts or road travel will drastically decrease. More likely both outcomes will occur, something that the UK government knows fine well.

One suspects this is why the country has been encouraged to switch to smart meters so that electricity can be cut off remotely in times of high demand. Proposals are afoot to allow power companies to declare emergencies that allow them to turn off electric vehicle chargers, home heating and other high power applications without compensation. This is a worrying prospect when we are dealing with greedy companies that exist solely to make money. It is even possible that one day, power companies could tap into power from electric vehicle batteries to prop up the grid. That would be a delightful surprise for drivers when they wake up to find their car battery flat on the day of an important meeting.

If these plans are being consulted on, we may as well take these scenarios as given. It is a sure bet that we *will* be suffering power cuts in the future because of an ideological madness driving policy that has questionable benefits, and likely no benefit at all, when it comes to climate change. Get a smart meter at your peril but you are better off sticking with analogue if you don't want your power cut off remotely. Longer term, we can probably count on smart meters becoming compulsory too, starting with rules for new builds. (What are the benefits of smart meters in any case – more accurate readings? A person can take his own accurate reading and submit that, so it's hardly even a benefit).

Another effect of electrifying the entire automobile fleet is that the government will lose billions in fuel taxes. Drivers of electric vehicles will not be purchasing fuel but they will still be using the roads, and so governments will seek to replace this lost revenue by replacing petrol taxes with something else. A flat fee on all electric vehicles would not be fair as the low mileage drivers would end up subsidising the high mileage drivers, so the likely solution will be a toll charged for every mile covered by electric cars. At first, these charges will be low because governments will want electric vehicles to appear to be an attractive option, but you can be sure the tolls will rise and rise.

It will also mean yet more state surveillance because the most efficient method for charging the toll will be with tracking devices in every vehicle, a standard feature of production in all vehicles before too long.

Whatever becomes of Greta Thunberg, and even if the entire media owned up to their biases and failure to report properly, every man, woman and child in the UK will have to live with this legislation unless a saner government repeals it. No one will say that the whole Greta Thunberg phenomenon was directly responsible for this Act of Parliament and indeed she is one cog in a very large machine. But every little helps, doesn't it?

The cost of implementing the UK's net-zero 2050 legislation has been estimated at a trillion pounds.[61] A trillion pounds and barely an eyelid was batted! This estimate included the UK Treasury's input so, if history is anything to go by, the cost will far exceed even this eye-watering number.

Whatever your views on Brexit, or indeed climate change, put them to one side for a moment. If you were unfortunate enough to witness the absolute ructions for several years about the £39 billion Brexit divorce bill, you must be wondering how something at least 25 times more costly could have slipped past with so little comment.

So yes, it does matter that the media conned us into thinking the Greta Thunberg circus was all-natural and wholesome. Here is a rule of thumb. If you agree with a particular cause but you still know that you are being lied to, do not be tempted to say, "Well, it's for a good cause, it's a noble lie" and just play along with it. A person willing to accept a lie, knowing it is a lie, is being willfully dishonest.

Imagine pointing out to somebody that the world is not going to end in 12 years nor will the world be broken beyond repair in 12 years at our current trajectory, as we are constantly told, whether the temperature is 1.5 degrees higher or not. If that person replies, "Oh yes, of course, I know that, but net carbon zero is still a noble cause and we should strive for it", that person is admitting they are being lied to but will not question the lie in order to be a good team player.

This is wrong. It is better to find out the truth, better to find out *why* they are deceiving you because the reason might shock you to your boots. In this case, it all has nothing to do with climate change, it is about replacing capitalism. Greta is a human shield to deflect attention from the billionaire class who are rolling out their Fourth Industrial Revolution and Great Reset. This is big news and it really ought to shock us more. If one just plays along with a nod and a smile, one is unlikely to ever find out the truth. It's not OK to play along for the 'greater good'. A campaign of manipulation on this scale is never for the 'greater good' and definitely not for your own good.

People should be angry when they are lied to by politicians and the media. It is disappointing that more people are not.

The other problem with 'being a good team player' and accepting a lie for the 'greater good' is that at some point down the road, that person is going to have to defend an issue in an argument that he knows is a lie. In doing so, he will have to lie himself. Accepting lies will turn an otherwise honest person into a dishonest one.

Accepting convenient lies risks storing up cognitive dissonance for later and things can get ugly when it comes to the inevitable day of reckoning. I have seen people become unhinged in front of my eyes when inconvenient facts are pointed out to them and they feel compelled to defend that cherished thing they know full well is a lie. They don't like it. Usually, they get angry and shoot the messenger. Refuse to accept the lies and you don't have to be that person.

Granted the charade is wearing thin by now, with Greta's 'How Dare You' tirade at the UN being something of a jumped-the-shark moment. However, in the beginning, only a minority could see through it.

Understandably, many people support Greta's cause because national governments, the entire media, the UN and countless other organisations, particularly the billionaire-funded NGOs, have spent decades telling us how important climate change is. People *want* to believe in the whole Greta phenomenon because they are wrapped up in the bigger narrative that climate change is the greatest threat to the planet. But to play along with the whole 'grassroots' meteoric rise aspect of the narrative is something else. Watching fully grown adults cheerleading Greta was cringe-worthy enough, but to see even non-mainstream journalists defend the phenomenon as a legitimate grassroots movement when they weren't being paid to hold such views, was a sorry sight indeed. It is one thing to say, "I can see through this charade but it's for a good cause so, I'll be a good team player" which is bad enough for the reasons just given. But to deny that the whole thing was being run from the top down by billionaires and through cynical media control is simply to deny reality. You see this even in the alternative media from people who really ought to know better.

What can we do? It's simple; don't get conned next time. In the words of Gerry Rafferty:

"… after a while, you get to recognise the signs

So if you get it wrong you'll get it right next time"

Granted, this is easier said than done but we all live and learn, or at least we should.

The first sign to recognise in this particular game is that whenever you see a person like Greta Thunberg, or an obscure psychology professor like Jordan Peterson, or a group such as Extinction Rebellion quickly elevated to superstardom with mass media coverage on multiple mainstream channels, you can be certain that rise was not organic. There is a reason for it. Always.

1.10

Bad Science

At this point, the reader is probably expecting a long section about the rights or wrongs of the global warming debate. Is it true? Will manmade greenhouse gas emissions cause mass extinction? Et cetera.

UN reports about climate change regularly run into four figures in page length, so let's try to keep this briefer. I had a fairly decent look at this issue to satisfy my curiosity over the last few years. That involved reading a pile of books, watching a few films, listening to podcasts, collecting hyperlinks and making copious notes. I was shocked at the rampant bias of the media, the outright fraud and dishonesty from people who should and no doubt *do* know better, the absence of the scientific method – usually from those who profess to follow and practically worship at the altar of 'the science' – and the politicisation of both governmental and non-governmental bodies, in particular the lamentable IPCC.

I am not a scientist and would never even try to pretend I understand how the climate changes over the decades, but one doesn't need to be a scientist to understand how propaganda works. One comes to learn that when the entire mainstream media, politicians, Hollywood stars, unindicted war criminals, the billionaire class and even royalty, are telling the same story, then it probably isn't true.

Take the orchestrated panic about the Amazon rainforest, for example. During August 2019, as fires raged in the Amazon, on practically every media channel we forever heard the refrain that the Amazon was the lungs of the planet. Superstars like Leonardo di Caprio tweeted that the lungs of the earth were in flames[62] and we were even told by the Secretary-General of the UN that the Amazon was a major source of oxygen.[63]

One might assume from this that the Amazon rainforest is the world's biggest oxygen producer. But it isn't true. Most of the world's oxygen is produced by the oceans.

The claim that the Amazon rainforest produces 20% of the world's oxygen is also false, a claim repeated by French President Emmanuel Macron,[64] and US Vice President Kamala Harris.[65] Macron also managed to slip in a comment that 'our house is burning', which one suspects he did not think of all by himself.

Our pants are on fire. LITERALLY.

The 20% is a big exaggeration with most estimates being somewhere between 6% (for instance James Foley, a climate scientist)[66] and 9% (Yadvinder Malhi, an ecosystem ecologist).[67]

But that is not the end of the story. Although trees emit oxygen, they also absorb oxygen. As *National Geographic* explains:[68]

"Trees don't just exhale oxygen – they also consume it in a process known as cellular respiration... So, during the night, when there's no sun around for photosynthesis, they're net absorbers of oxygen. Malhi's research team reckons that trees inhale a little over half the oxygen they produce this way. The rest is probably used up by the countless microbes that live in the Amazon, which inhale oxygen to break down the dead organic matter of the forest.

'The net [oxygen] effect of the Amazon, or really any other biome, is around zero,' he explains."

Did you catch that? The net oxygen effect of the Amazon is around zero! Forest fires in the Amazon are not depleting the world's oxygen supply. This is not to say that fires in the Amazon are a good thing or that one does not care about the environment and by default hates small children and kittens. It is just to point out the dishonesty and emotional blackmail of it all.

The Amazon suffers from forest fires every year, although politicians only started boohooing about them recently; and lying. It seems that there is something about fires that sets off the urge in certain people to tell whoppers about climate change. As the old proverb doesn't go: 'when you see a forest fire, you won't wait long for a climate change liar'.

This example is only a minor quibble in the scheme of things. Most people who bleat that the 'Amazon rainforest is the lungs of the planet' are just innocently repeating misinformation rather than deliberately setting out to deceive. They believe it because they heard everyone else say it. It is inexcusable for professional media outlets to fall victim to this behaviour though, especially when Presidents, Vice Presidents and the UN Secretary-General are repeating the claim. Their job is to check these things.

1.11

Ocean Health

We are forever being told that the coral reefs are dying because of coral bleaching, which occurs when seawater is warmed by hotter temperatures.[69] We are warned that reefs are in danger of disappearing altogether within a century,[70] before being told they are making miraculous recoveries[71] and even returning from the dead,[72] which was not supposed to be possible. Does this fill you with any confidence that the alarmists know what they are talking about?

We are also constantly told that islands such as the Maldives, Tuvalu and Kiribati are in danger of slipping underwater as sea levels rise. We've all seen headlines like these:

'Climate change threatens Maldives' (*CBS News*)[73]

'Tuvalu – One day we'll disappear' (*The Guardian*)[74]

'Kiribati – life on a tiny island threatened by the rising sea' (*The Guardian*)[75]

Despite the dire warnings these islands refuse to drown. The positively eye-watering property prices in the Maldives – the canary in the coal mine for global warming alarmists – are not exactly indicative of an incipient disaster zone.

Not only are these supposedly threatened islands not sinking, but many of them are also actually *growing* in size. This was demonstrated by coastal geomorphologist Paul Kench who has studied how reef islands in the Pacific and Indian Oceans respond to rising sea levels:[76]

"Here we present analysis of shoreline change in all 101 islands in the Pacific atoll nation of Tuvalu. Using remotely sensed data, change is analysed over the past four decades, a period when local sea level has risen at twice the global average ($\sim 3.90 \pm 0.4$ mm.yr^{-1}). Results highlight a net increase in land area in Tuvalu of 73.5 ha (2.9%), despite sea-level rise, and land area increase in eight of nine atolls. Island change has lacked uniformity with 74% increasing and 27% decreasing in size."

Reality check: three-quarters of Tuvalu's islands *increased* in size over the past 40 years. This is a fact. It has been known for years that reef islands are not shrinking in general and it is the height of dishonesty to keep crying wolf.[77]

The climate alarmists were wrong. This is also a fact – one they should admit to and reassess their views of before repeating the same old disproven conjecture. Some other factor than a sea-level rise is at play here or else the islands could not grow in size. An honest mind would try to find out what that was.

Kench has a good explanation of why these islands 'bounce back', so to speak:

"Existing paradigms are based on flawed assumptions that islands are static landforms, which will simply drown as the sea level rises. There is growing evidence that islands are geologically dynamic features that will adjust to changing sea level and climatic conditions."

Reef islands are some of the most dynamic landforms on earth. They move in response to shifting sediments. They are like construction sites, the reefs acting as factories producing a constant supply of raw building material, namely calcium carbonate. As Ugo Bardi, teacher of physical chemistry at the University of Florence, explains:[78]

"The reason is that the islands, or at least the reef barriers around them, are *alive.* They are not just chunks of rock emerging out of the ocean surface. They are the result of the mineral excreta of tiny creatures that create the hard part of the coral barrier with their exoskeleton.

Some coral reefs survived the great sea level rise (some 120 meters!) that took place at the end of the last ice age. Not a small feat, but it was possible over a few thousands of years."

That last amazing point highlights the fact that reef islands can adjust to their surrounding habitat, in this case rising sea levels. Theoretically, sea levels might rise faster than the reefs' ability to adjust, which might be cause for alarm if this was the case. Indeed the crux of global warming panic revolves around the fear that ice caps will melt and the ocean levels will rise, leading to flooding. But how fast are sea levels rising? This is another area that needs addressing.

Paul Kench's study above referred to sea level rises of 3.90 mm per year.[76] NASA has estimated a recent increase of 3.3 mm per year,[79] which represents faster than historical increases. A different report from NASA estimates sea levels rose 160-210 mm in the last century.[80] The Royal Society says 3.6 mm per year and states that the overall observed rise since 1902 is about 16 cm (6 inches).[81]

Historically, we are talking sea level rises of less than one foot in the past century and if we were to say two feet in the next century we would probably be being unduly alarmist.

But that doesn't stop the scaremongering *Guardian* from claiming that sea level rises not only *could* be as high as twenty feet in the next century but *will be*. The *Guardian* must be employing clairvoyants these days (actually a geography professor, Harold Wanless) with headlines like this one from April 2021:[82]

"Sea levels are going to rise by at least 20 ft. We can do something about it"

The reason for this estimate? The sub-headline contains both the alleged answer and the 'solution':

"To avoid the grimmest outlook posed by warming oceans, we need to extract heat-trapping gases from the atmosphere."

Quel surprise. Someone at *The Guardian* wants to promote ferociously expensive carbon capture schemes (CCS) whose efficacy is unproven. The author continues:

"But if seas rise 20 feet or more over the next 100 to 200 years – which is our current trajectory – the outlook is grim. In that scenario, there could be two feet of sea-level rise by 2040, three feet by 2050, and much more to come."

There is no evidence to suggest that this will happen. There is no reason to believe it even *might* happen, based on literally anything, least of all the report that *The Guardian* hyperlinks in its own article. The report that readers are steered to was written by NOAA (the USA's National Oceanographic and Atmospheric Administration).[83]

It is quickly apparent that there is nothing in this report to suggest that 'our current trajectory' will lead to anywhere like a 20-foot rise. Estimates in the graph on page 12 of the report were between 20 cm and 200 cm, which is only 6 feet.

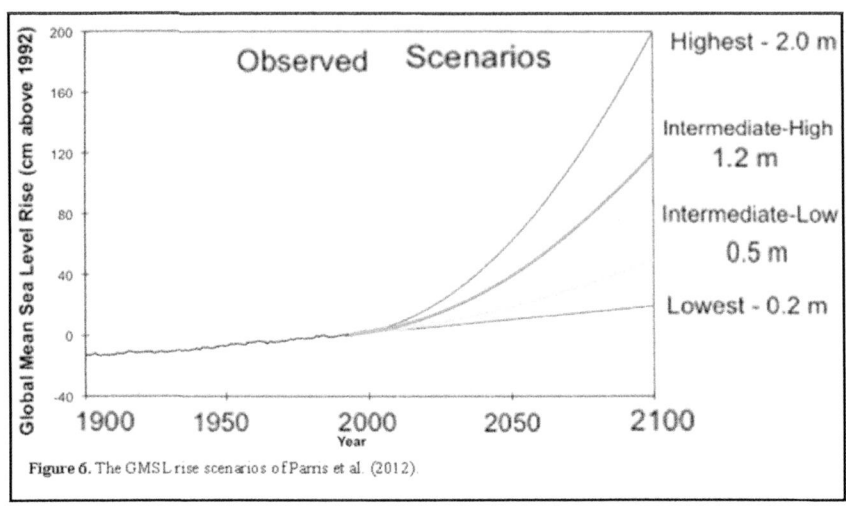

Figure 6. The GMSL rise scenarios of Parris et al. (2012).

In fact, the 'high end' estimates quoted in that report under the heading future sea levels: scenarios and probabilistic projections varied between 1.8 m and an absolute worst case 2.5 m, as indicated by this quote:

"... estimates of high-end GMSL rise by 2100 under RCP8.5 include ~1.8 m [95th percentile] (Jevrejeva et al., 2014, Rohling et al., 2013 and Grinsted et al., 2015), ~2.2 m [99th percentile] (Jackson and Jevrejeva, 2016), and ~2.5 m [99.9th percentile] (Kopp et al., 2014)." (Page 13 of the report)."

An absolute worst case, extreme, most pessimistic scenario of all was described on page 14 to mark an upper bound for the report:

"... this report recommends a revised worst-case (Extreme) GMSL rise scenario of 2.5 m by 2100".

How *The Guardian* translated a worst-case scenario of 2.5 m by 2100 to 'sea levels *will* rise 20 feet at our current trajectory' (with an added 'at least') is anyone's guess. Here is a diagram to demonstrate the size of this lie. You might need to squint to see the original graph.

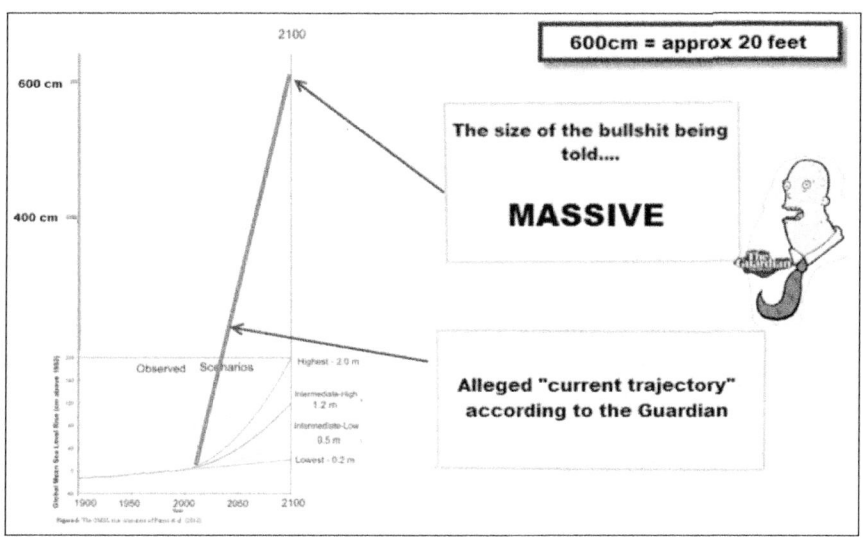

The Guardian's own source completely refutes its crackpot claim that sea levels will rise by 20 feet in the next century or two. Self-refuting journalism at its best.

As to our 'current trajectory' being of that magnitude, that is laughable and the figure appears to be plucked out of thin air.

This is pure fear-mongering. It is shameful, dishonest journalism.

Sea levels can rise by hundreds of feet over tens of thousands of years, but this sort of thing happens at the end of an ice age. Right now, a foot per century seems a lot more likely. Even the IPCC's 4th Assessment Report of 2007 predicted a sea-level rise of a maximum of 59 cm in its most pessimistic case.[84]

There is a human activity far more dangerous to the more popular islands in the Maldives than fossil fuel burning and eating beef: the building of new runways, roads and large buildings. These structures are extremely heavy and are locked in place. Gravity counters the action of the reefs to stay afloat. Ceasing to build runways and hotels would be a more advisable policy than to stop driving, eating red meat or imposing carbon taxes. The latter will make absolutely no difference to whether the Maldives ends up underwater or not.

Who can be sure though? Perhaps nature will win out over even this building activity as well?

1.12

When 97% 'Consensus' = 100% Bullshit

One of the most outrageous climate change claims ever made is that 97% of scientists agree that climate change is manmade. This statement is a total falsehood. Even establishment media outlets like Forbes have admitted this, for example with a 2015 headline:[85]
"'97% Of Climate Scientists Agree' is 100% Wrong"

The claim is so far from the truth it has to rank as one of the biggest lies emanating from climate alarmists' camp and there is some pretty stiff competition for that honour. The genesis of the 97% claim is a 2013 survey led by Jonathan Cook. In this survey of around 12,000 papers, two-thirds of the authors did not even express an opinion as to whether climate change was manmade. The 97% claim is complete bunk. In Cook's words:[86]

"A new survey of over 12,000 peer-reviewed climate science papers by our citizen science team at Skeptical Science has found a 97% consensus among papers taking a position on the cause of global warming in the peer-reviewed literature that humans are responsible."[87]

To be fair to Cook, he did not claim '97% of all scientists agree climate change is manmade'. It tends to be the media, politicians and unthinking drones who like to repeat what they hear on the news that made this claim. Cook was referring to the scientists who were 'taking a position'.

However, this isn't true either. Several authors have flat out refuted the categorisation that Cook's team assigned to their paper, some of them irked at the reviewers' insinuation they knew better what the authors were thinking than the authors themselves.[88]

As Cook et al's paper admits, they were only 'examining 11,944 climate abstracts from 1991–2011'.

By only examining the abstracts and not the papers themselves, one might charitably attribute the falsely classified papers to incompetence rather than bias or dishonesty. What is certain is that the conclusions are baseless and the attention given to the Cook survey by the media and politicians, unwarranted. It is almost as if certain people are trying to sell a narrative to us.

Many people believe the 97% claim because they hear it repeated in the media and by silver-tongued liars like Barack Obama. Believing they know the 'facts' they might be less inclined to study the issue, which is probably the whole point.

To those wondering what is the real percentage of climate scientists who believe climate change is manmade, the truth is very surprising. An analysis by Mark Bahner found that although 32% of authors implicitly or explicitly agreed man's activity has *some* effect on the climate, only 64 papers explicitly claimed that man was the *main cause* of climate change.[89]

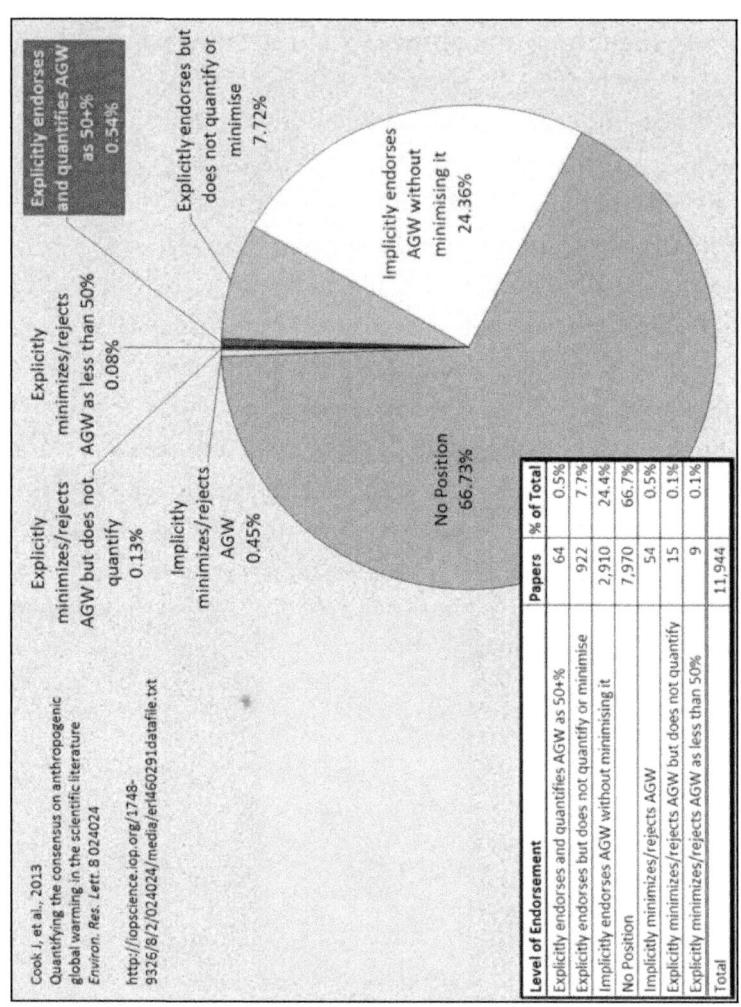

A far cry from 97%.

By 'main cause' the criterion that at least 50% of climate change is manmade was applied. Considering the endless hysteria suggesting that man is literally the only cause, this criterion is not unduly onerous. Only 64 papers out of 11,944 fit this criterion. Given that 7,970 of the 11,944 papers expressed no view, 64 out of the remaining 3,974 papers where the author did express a view is a paltry 1.6%. Quite some way from 97%.

Would the BBC only deem these 64 authors to be worthy of a television interview and decry everyone else as a 'denier'? Former IPCC lead author Dr. Richard Tol criticised the media for their false claims about the alleged 97% consensus and said that Cook's analysis was a load of old bollocks.[90]

In May 2014, Tol testified to the US Congress that the 97% was essentially pulled from thin air, it was not based on any credible research whatsoever.[91] People still harbouring delusions that the 97% claim might be legitimate, should have a look at an online article titled: *97 Articles Refuting the 97% Consensus* which completely and utterly dismantles Cook's badly designed and executed survey. Then they can realise for themselves what a joke the 97% claim is.[92] Jonathan Cook still tries to claim his study has value, with an article on his website titled *The Cook et al. (2013) 97% Consensus Result is Robust*.[93]

Bless him. Jonathan Cook calls his echo chamber, sorry I mean his website 'skepticalscience.com', which is so laughable it could be a case for Trading Standards. It is the antithesis of 'sceptical' (British spelling please), with honest debate snuffed out in the heavily moderated comments section when site administrators fear they are losing control of the narrative. Try it yourself. Comments that do not meet the approval of the global warming camp are removed so rapidly it isn't too far from real-time editing.

Compare this behaviour to that with genuine sceptic websites such as realclimatescience.com or wattsupwiththat.com. Free debate is permitted in the latter two 'heretic' sites and opposing comments are debated in good faith, usually ending with the alarmist beating a retreat after politely having his arguments dismantled.

If '97% of scientists agree' and the climate change sceptics were so wrong, it ought to be a simple matter to *demonstrate* that they are wrong with a well-formed argument. Why run and hide? Probably for the same reason, the green lobby uses Greta Thunberg as a human shield and the BBC is not prepared to give sceptics air time. They know their case is that flimsy they would lose the argument. A true sceptic or real scientist is not frightened to debate honestly and does not shut down the debate by silencing his critics. Only cowards and people who have something to hide act in this fashion.

One doesn't need to know the first thing about science to see these people for what they are.

1.13

The C Word

As to the 97% consensus what on earth is 'consensus' supposed to count for in a scientific debate in any case?

Science is not conducted by a show of hands. It is not a democratic process. At the time Galileo believed the earth revolved around the sun, he was pretty much out there on his own and he was in the right. It is said that in response to the book *100 authors against Einstein*, Albert Einstein remarked, "If I were wrong, it would only take one." The story is probably apocryphal but the sentiment is valid. In science, your theory is either correct or it is not. You do not win the argument with a vote.

As well as endless appeals to authority ('trust the science' guys) we are subjected to these appeals to herd instinct and told there is 'consensus' on the matter. It's an old trick that propagandists use. To convince people to believe in something just persuade them that everyone else believes it. Most people won't bother to investigate the issue and hate being part of a minority, especially a minority that is subject to public scorn or ridicule, so they just take the path of least resistance and conform.

From time to time, we see well publicised exercises along the same lines as the *100 authors against Einstein* book. A letter will be passed around signed by a number of important people and much airtime given to the existence of the letter.

In November 2019, one such exercise backfired after a letter gathered 11,000 signatures from scientists declaring a climate emergency. After plenty of hand wringing headlines like *The Guardian's* 'Climate crisis: 11,000 scientists warn of 'untold suffering',[94] the letter's gravitas took a hit when someone spotted the signature, among many other fakes, of 'Professor Micky Mouse' (sic) of the 'Micky Mouse Institute for the society of the Blind, Namibia'.[95]

Anyone could sign this letter and clearly did, including a Professor Dumbledore from Hogwarts and one Araminta Aardvark from the University of Neasden, but a headline screaming '11,000 randoms sign petition' probably doesn't carry the same weight as a warning of 'untold suffering' from 11,000 scientists.[96]

There is another well-known petition sponsored by the Oregon Institute of Science and Medicine (OISM). As far back as 2008, it had been signed by more than 31,000 scientists, over 9000 of whom held PhDs.[97] The petition states:

"There is no convincing scientific evidence that human release of carbon dioxide, methane, or other greenhouse gasses is causing or will, in the foreseeable future, cause catastrophic heating of the Earth's atmosphere and disruption of the Earth's climate."

Now, I can't vouch for the scientists' credentials on this list either, but some of them have to be real. Certainly enough to show there is no 'consensus' on the issue. How about this headline, on a US government website no less? The US Senate Committee on Environment and Public Works published a report with the headline: *U.S. Senate Report: Over 400 Prominent Scientists Disputed Man-Made Global Warming Claims in 2007.* The sub-headline even spells it out: *Senate Report Debunks 'Consensus'*.[98]

The alleged scientific consensus on climate change is a great big lie propagated by the climate change lobby, extremely damaging to the truth and perhaps the one that the lobby has leveraged the most capital out of.

There is no consensus on the issue of manmade global warming. The science is not settled. These are simple statements of fact. Anyone who says it is, or who repeats the 97% claim is either lying or hopelessly ignorant. When somebody argues that '97% of scientists agree and I'm on their side' as if scientists were infallible and let's all make a big appeal to authority, you are probably dealing with a person who has not even spent one hour investigating the issue. In my personal experience, often these are the individuals who have that awful mixture of ignorance and smugness (the dreaded 'smugnorant' type), who speak in condescending tones as though you were the one not getting it. One hopes that if they knew how badly they had been deceived, they might tone it down a little but humility is not what these people tend to offer. In these situations, try asking them what the letters IPCC stand for. They are almost guaranteed not to know.

While we are on the subject of facts here are a few more:
- evidence of climate change is not evidence of anthropogenic global warming (AGW),
- evidence of global warming is not evidence of AGW,
- evidence of rising CO_2 is not evidence of AGW.

In my experience, it is not a fruitful exercise to argue this topic with people. Most of those who have decided one way or the other on this issue will not change their mind regardless of the evidence because they do not care for evidence. Many adherents to the manmade global warming hypothesis believe in it as people believe in religion. I have also found that disciples of the AGW religion tend not to debate honestly or in good faith. Facts do not matter to them.

Suffice it to say, the concerning problem facing humanity is not that burning fossil fuels will heat the planet to the point of destruction. If anything, the bigger problem facing humanity is that one day we will run out of fossil fuels to burn.

This brings us to the concept of 'EROI'. EROI is the issue that 99% of economists have failed to grasp, probably because it is occulted from economics classes by those who realise the importance of the concept. People don't talk about EROI on the news either, even on the financial channels. If everyone understood EROI, more of us would understand why the people who run this world are hell-bent on a 'reset'.

EROI stands for 'Energy Return on Investment'. Take the following scenario:

Imagine you discovered 100 billion dollars' worth of oil. Happy days! One problem – it is located 100 miles below the surface of the earth. Now, it is going to cost over 100 billion dollars to get it out of the ground. It is a downright impossible task with existing technology, so you don't bother.

It takes energy to get oil out of the ground. Back in the 1900s, one could drill for oil and hit a rich seam, perhaps getting 80 or even 100 barrels of oil for each barrel you used to get it out of the ground, or even more if it was a massive oil field. EROI is also referred to as EROEI, which stands for Energy Returned on Energy Invested. We use the terms interchangeably, but EROEI is a more accurate name as it illustrates the like for like *energy* comparison we are making. Today most of the low-hanging fruit has been picked, so to speak, so the EROI has steadily fallen since the 1900s. There is oil out at sea but it is tougher to extract from under the sea bed. So perhaps EROI is 10:1 for each barrel invested. As the more preferable opportunities at sea are used up, EROI falls further. Some say current rates of EROI are as low as 5:1.

You can see where this is heading. At some point you fall below an EROI of 1:1 and fossil fuels will be more expensive to extract than they are worth. At this point, they will just remain in the ground and people will not bother to extract them. Alternatively, the industry will sell it at a loss and the government will bail out their friends at the oil companies and give them taxpayer money to drag oil out of the ground at uneconomic rates. (This was my attempt at a joke by the way, although you never know with the lunatics in power.)

Much of the US shale oil industry has already reached the point where it is uneconomic to get oil out of the ground. The US shale oil industry is financed with so much debt it is in effect a massive Ponzi scheme and we all know what happens to Ponzi schemes when new suckers can't be found to finance them. It is an industry about to crash and the share price of almost every specialist US shale oil company is on the road to zero in the next decade.

Either we run out of fossil fuels to burn, or we do not run out but they stay in the ground untouched. Same difference. Humanity needs an alternative to fossil fuels. What about so-called sustainable energy, namely wind and solar?

Let us nip this in the bud. It can be said with absolute certainty that humanity cannot hope to maintain its current standard of living with wind and solar power. These ideas are environmental pie in the sky and completely unattainable in their current forms.

Even Michael Moore admitted this in his 2019 film, Planet of the Humans, something that drew the ire of many an eco-warrior and even his liberal acolytes for going off message. But Moore was right. The main 'sustainable' green energy solutions such as wind and solar are nowhere near sustainable and cannot hope to be provided anywhere near the scale that is required. They are also far from 'green'. Wind turbines require huge amounts of cement, steel and therefore fossil fuels to erect and roads are often cut through forests to deliver them to their destination. Their operation slaughters thousands of large birds unfortunate to cross their paths and the turbine blades are thrown into landfills after 15-20 years of operation, many before even that long. Solar panels also require a massive fossil fuel input and silver needs to be mined to create the panels, a dirty business if ever there was one. If the panels lasted forever that would be one thing, but they only last for 25 years while efficiency declines throughout their lifetimes. Sorry to put it so bluntly, but when you research this for yourself you will come to the same conclusion: wind and solar power are duds. They might complement the status quo but they will never be able to replace it.

Talk of achieving a 100% renewable energy supply is delusional. Achieving even fractions of that goal is an impossible dream because it requires more resources than the planet can provide, particularly with regards the supply of rare earth metals. Anyone can verify this for themselves. Electrifying all the world's cars and running them renewably, for instance, would require us either to mine asteroids or to find a few spare planet earths somewhere to make up the supply. It just ain't happening.

And that's just to speak of motor vehicles. Shipping requires an even greater amount of energy and that means liquid fuels in large amounts, ie derivatives of oil. Without shipping, world trade collapses.

Now ask yourself this: do you think the people pushing the Green New Deal and the politicians charged with implementing these policies, are unaware of all this? Are they incapable of doing the internet search that informs them about the world's cobalt, nickel, lithium and neodymium supplies?

Of course not! They know fine well these policies are physically impossible to bring about with a world population of almost eight billion, yet they press ahead regardless. The only possible way these plans could ever come to fruition is if the population were to be vastly reduced. Could this be the real objective? After all, these policies are guaranteed to cause untold poverty and death, while offering next to zero environmental benefits.

Throwing money at wind and solar benefits the companies that provide them and that is about it. Granting taxpayer subsidies to these industries only serves to impoverish the unfortunate taxpayers of those countries. When Greta and the rest of the environmental movement encourage a speedy transition to expensive sustainable green energies that aren't sustainable or green, they are only hastening this impoverishment. Another dishonest trick the green lobby uses is to claim renewable energy is competitively priced, neglecting to mention that wind and solar benefit from massive subsidies while oil, for consumers, has always been taxed to the hilt.

The obsession with 'net-zero' and the war on carbon dioxide as some kind of poison will end up causing immense human misery at this rate. One incident stood out that demonstrated both Greta Thunberg's lack of understanding of this issue and the willing connivance of the media at the same time. One day, when Greta was giving a speech at Davos (where else) she exhorted the world to strive for *actual* zero carbon emissions, not just 'net' zero.[99] This was a staggering pronouncement and just as staggering was the fact it went without challenge.

Anyone who knows anything about these matters can explain how *actual* zero CO_2 emissions for a decent length of time – not just 'net' zero CO_2 emissions – would lead to the death of hundreds of millions if not billions of people. They would die of cold and they would die of starvation. Actual zero would not only mean no cars, but it would also mean no planes, no gas, no oil, no heating, no industry, farming crippled and massive food shortages in rapid order, starting in the cities where trouble would be guaranteed to break out. Chaos would ensue, after which the vast majority of us, excluding a tiny group of people fortunate enough to have their own off-grid power systems, would revert to burning wood to keep warm.

Did Greta wander off script there? If she did the instant response that ought to have been forthcoming if we had anything like a competent and honest media did not materialise.

Colder climates are far more deadly to humans than warmer ones and the policies that the green lobby agitates for will cause millions to realise this brutal reality. A 2015 study that analysed 74 million deaths in 13 countries between 1985 and 2012, found that 5.4 million deaths were related to cold, while only 311,000 were related to heat.[100] That is almost 20 times the difference. Cold weather harms the human body more than hot weather, in particular the respiratory and cardiovascular systems (so you wouldn't want to have COVID-19). There is a good reason that the homeless gravitate to hotter places like Las Vegas and San Francisco.

Expect to see more images like this one in first world countries as a result of environmental policies.

Forget Greta's actual zero. Merely net-zero will cause death and suffering from cold as millions of people will no longer afford to be able to pay their heating bills. The inability to travel long distances in electric cars when petrol cars are phased out will also lead to a loss of independence and opportunity. The scandalous thing is that policy makers understand this fully, while Greta obviously does not or else she would not call for absolute zero CO_2 emissions. Greta

Thunberg might believe in her mind that actual zero CO_2 is a desirable outcome but this only serves to demonstrate her ignorance on the matter. This is sad, for she seems almost to have been programmed to believe that carbon dioxide is evil. Someone needs to sit her down and explain to her that plants, animals and humans all need CO_2, that it is a vital building block of all life on earth and not her enemy.

1.14

Is That All?

The amount of manmade CO_2 from cars, planes, ships, factories, dropping bombs and so forth only contributes a tiny fraction of the total CO_2 in the atmosphere. One 2017 study estimated the manmade figure to be 4.3%.[101]

A different 2017 estimate from the European-based CO_2 Human Emissions Project stated that: "Our output of 29 gigatons of CO_2 is tiny compared to the 750 gigatons moving through the carbon cycle each year", which equates to a manmade percentage of 3.87% of the total.[102]

One can argue whether this number is 3%, 4% or 5% but the point holds; the vast majority of CO_2 in the atmosphere is not manmade but naturally occurring, most of it being released from the oceans over time. The oceans hold more than 50 times the amount of CO_2 that exists in the atmosphere.[103]

When one grasps what a tiny fraction of the atmosphere CO_2 represents to begin with, the 3-5% manmade portion looks even more innocuous. By volume, dry air contains just 0.04% carbon dioxide. We are talking about a gas that forms around 400 parts per million of the atmosphere *to begin with*, and then taking 3-5% of those 400 parts! That would be 12-20 parts per million.

For the record, the whole atmosphere consists of 78.08% nitrogen, 20.95% oxygen, 0.93% argon, 0.04% carbon dioxide, and small amounts of other gases (neon 0.001818%, methane 0.000524%, helium 0.000187%, krypton 0.000114%).[104]

Would you care to see what this looks like in a picture? Even if we are generous to the warmists and accumulate what remains of the two centuries worth of manmade CO_2 it still only amounts to the very final, single dot in the image below.

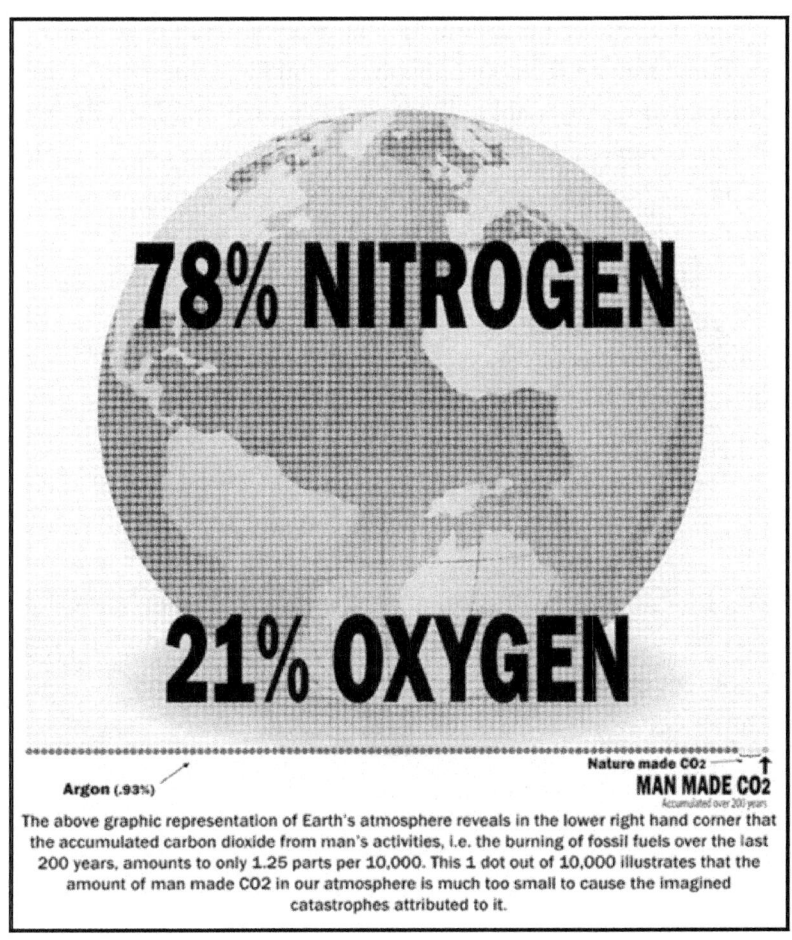

The above graphic representation of Earth's atmosphere reveals in the lower right hand corner that the accumulated carbon dioxide from man's activities, i.e. the burning of fossil fuels over the last 200 years, amounts to only 1.25 parts per 10,000. This 1 dot out of 10,000 illustrates that the amount of man made CO2 in our atmosphere is much too small to cause the imagined catastrophes attributed to it.

Are we really to believe that the human-created CO_2 represented by that single dot *drives* climate change? What about the 380 parts per million that are not manmade?

Here is a fact of life: whether CO_2 is produced by humans burning fossil fuels or whether it occurs naturally, it is still CO_2. Rest assured Mother Nature does not distinguish between the two. So even if we did reduce manmade CO_2 to zero, eventually killing hundreds of millions/billions of humans through starvation and cold, the world would still be left with around 95% of the total amount of CO_2 it started with. What a pointless endeavour that would be!

The fact that Greta's 'absolute zero' statement was not thoroughly examined, ripped apart and honestly dealt with, suggests that there is no appetite for an honest examination of the issue. It is as clear as day that an awful lot of people are simply not interested in the truth or breaking ranks with the narrative.

If you have not heard this next little fact, it is certain to surprise you. The amount of CO_2 in the atmosphere *increased* during the first lockdowns in 2020. The May 2020 reading of CO_2 at the Mauna Loa Observatory was registered at 417.1 parts per million, the highest level in human history.[105]

How does a manmade global warming alarmist possibly explain this? Do you remember throughout the lockdown, all those articles saying how wonderful it was that our lovely blue skies were back and that the lack of pollution was cleaning up the earth? It was portrayed as perhaps one of the benefits of lockdown (almost as though someone was pushing an agenda).

We have constantly been told that the amount of CO_2 is rising in the atmosphere because of man's dirty industrial usage. Between March and May 2020, the rates of CO_2 emitted from cars, planes, factories and other industrial sources would have fallen off a cliff. Most businesses were shut and the roads and skies were practically empty for quite a time.

Yet, CO_2 levels in the atmosphere increased!

It might have only gone up by a few parts per million, but according to the warm mongers it had no business increasing at all with half the world shut down. Scientists have attributed this rise to 'natural variability', to which climate realists around the world replied:

1.15

Water Deniers!

For all the obsession about carbon emissions, CO_2 isn't even the most abundant greenhouse gas in the atmosphere. When we listed the components of the atmosphere above (nitrogen, oxygen, etc) it was with the proviso that the air was dry.

In other words, this was ignoring humidity. This is not a reflection of reality: water vapour can vary locally from anything around 0.1% in cold areas to as much as 4% of the atmosphere in tropical areas (in which case the other gases' percentages are scaled-down proportionately).

Water vapour is by far the most important and abundant greenhouse gas in the atmosphere, dwarfing the proportion of CO_2. If water vapour is 0.4% in a particular area, it is ten times more abundant than CO_2. If water vapour is 2% then it would be 50 times more abundant than CO_2. Yet it is practically ignored in discussions about climate change.

Now, CO_2 and H_2O vapour are not identical gases and they do not have the same effects in the atmosphere. However, there is no denying that H_2O has the far greater greenhouse effect and it is peculiar that this issue receives so little attention.

Try an internet search for 'greenhouse gases pie charts' and you will see the vast majority of the charts completely ignore water vapour! It is as if we are only supposed to focus on manmade greenhouse gases. Here is a pie chart that did manage to include the most abundant greenhouse gas, our good friend H_2O (on the left in the image below).[106]

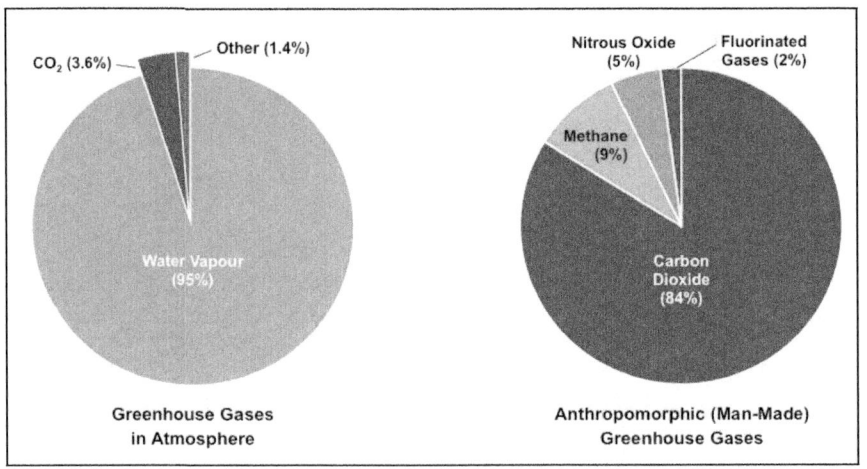

Right beside is a second pie chart showing purely manmade greenhouse gases. Can you see the difference between an alleged SIGNIFICANT human contribution to the greenhouse effect and the reality of a NEGLIGIBLE one?

When you look at the chart on the left and consider that man only contributes less than a twentieth of the 3.6% slice of CO_2, one comes to understand that the real crackpots in this debate are the people who claim that climate change is only caused by man.

The problem for the climate alarmists is that the quantity of water vapour in the atmosphere is independent of human activity.[107] Some say that 99.999% or more of water vapour is of natural origin.[108]

We could stop the planes, trains and automobiles, close all the factories and blah blah, we could even drink a glass of kool-aid each and commit mass planetary suicide until not a single human being remained, and not only would the world be left with 95%+ of the carbon dioxide it started with, it would still have 99.999% of the water vapour as well.

Water vapour is deliberately excluded from the conversation because it puts the whole climate change scare into a clearer perspective. Including water vapour in the equation would lessen the fear of climate change because people would realise other factors play a bigger role in the natural warming cycle of the Earth. People would rightly ask: you want me to stop flying, driving and eating meat for *that*?

The climate lobby knows that water vapour is more important than CO_2 and methane, but there is absolutely nothing we can do to reduce water vapour so they skirt over the issue. Perhaps it is time to give the alarmists a taste of their own medicine and label them 'water vapour deniers'.

But who knows, perhaps some genius might devise a massive planet-sized dehumidifier to hover in space and suck all the evil water vapour out of the air to protect us from ourselves. Who needs rain anyway?

Would this be any less insane than trying to suck carbon dioxide out of the atmosphere on immense scales and at a mind-boggling cost? Just watch as they attempt to try it, knowing that the costs will be falling on you.

1.16

Inconsiderate Temperature Data

Next, there is the 'problem' of the average world temperature not increasing as it was supposed to. After around 1998, the global average temperature failed to keep rising, despite the climate change lobby insisting that increased CO_2 levels would drive up temperatures. CO_2 levels in the atmosphere have indeed risen steadily for decades, yet the average global temperature could not keep up the pace.

You can imagine what a huge problem the hiatus in warming caused for the climate change lobby. Every possible reason and excuse has been proffered to explain how this does not invalidate the beloved theory of manmade climate change, with 66 such excuses being documented on one website alone.[109]

Here is one such attempt to explain away the pause. On 23 September 2013, the BBC reported that the pause would be central to the next IPCC report, which was yet to be released.[110] The article stated: "Many governments are demanding a clearer explanation of the slowdown in temperature increases since 1998. One participant told BBC News that this pause will be a 'central piece' of the summary."

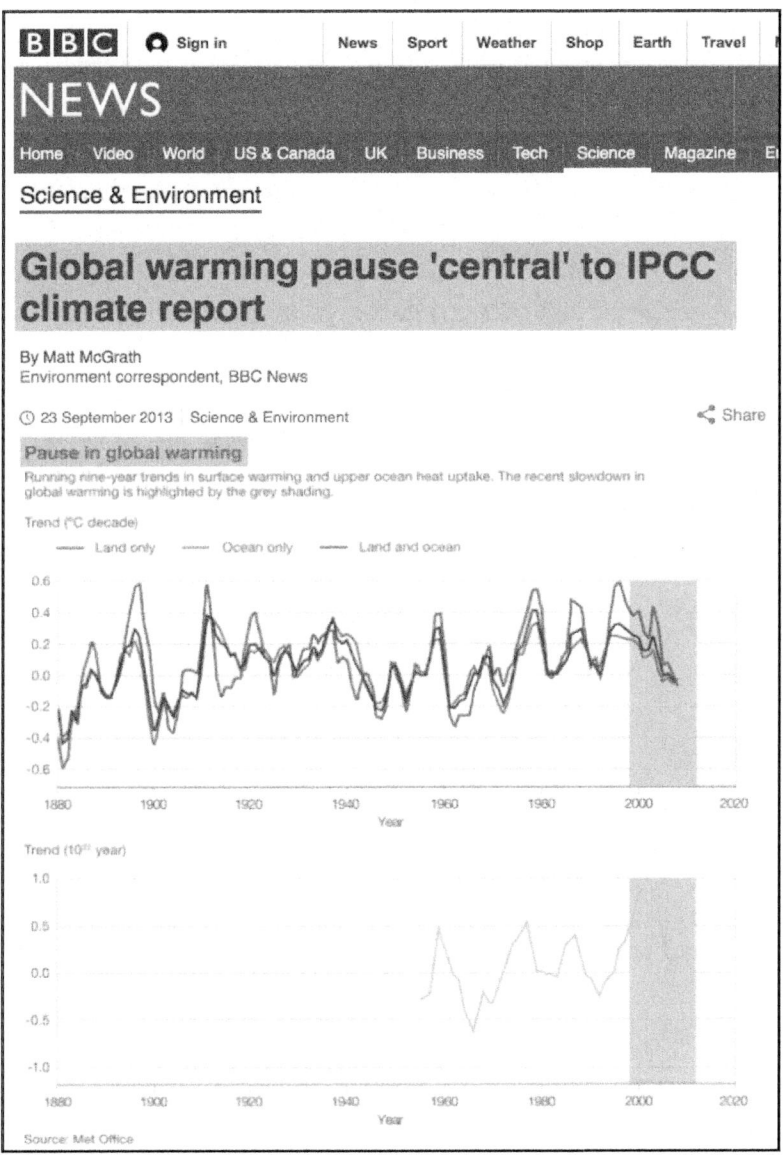

How would the IPCC explain the pause? Four days later on 27 September 2013, *The Guardian* gave its interpretation of the IPCC's report for policy makers.[111]

They declared the pause to be a 'mirage' along with an image of a polar bear standing on thin ice.

Images of polar bears have been used again and again as a manipulative device to tug at readers' heartstrings and make people feel guilty for climate change. Indeed it was this exact kind of image that so upset Greta Thunberg it made her ill and inspired her into activism. *National Geographic* apologised for a misleading 2017 report on an emaciated polar bear that went viral and it won't be the last time this happens.[112] People need to understand that when we see a starving polar bear on a melting piece of ice it is not manmade climate change that caused it. It is a fraudulent narrative, as is the narrative that polar bear numbers are decreasing. Polar bear numbers have been *increasing* for years and the species is not under threat.[113] In fact, the polar bear population is in rude health, a fact documented superbly by leading expert Susan Crockford at the website polarbearscience.com.[114]

This ought to be good news for those who are worried about climate change but strangely this information is ignored. Instead of praise for her fine work, Susan Crockford has found herself the subject of snide commentary online,[115] maligned on Wikipedia [116] and she lost her position as an adjunct professor. After 15 years, her contract was not renewed at the Canadian University of Victoria. She describes this as 'an academic hanging without a trial, conducted behind closed doors.'[117]

Having set its stall out with the polar bear, *The Guardian* mocked the 'idiots' who ever doubted the climate scientists with the sub headline:[111]

"Only 1% of the heat trapped by greenhouse gases warms the air, making the pause claimed by IPCC critics an idiotic sideshow."

Er, right, but that doesn't explain anything, least of all why it is 'idiotic' to talk about the pause. Soon enough, *The Guardian* began to preach the same old message as ever: it's human activity, stupid.

"The landmark new report from the Intergovernmental Panel on Climate Change (IPCC) is crystal clear: human action is warming the planet and we're heading for big trouble if carbon emissions are not slashed...

Yet, before the ink is even dry, critics are trying to obscure this stark message behind a mirage: the supposed halt in global warming over the last 15 years. This willful idiocy is based on the fact that air temperatures at the Earth's surface have more or less plateaued since the record hot year in 1998.

What critics choose to ignore is that of all the extra heat being trapped by our greenhouse gas emissions – equivalent to four Hiroshima nuclear bombs every second – just 1% ends up warming the air. By choosing to focus on air temperatures, critics are ignoring 99% of the problem."

There you have it. A *Guardian* word salad even its author would struggle to explain. As to the 99% of the heat that *The Guardian* claims is not warming the air, what is it – fake heat?

What *The Guardian* is asking us to believe is that temperature readings no longer matter. Temperature, the whole crux of the global warming argument for years is not important and pointing out the pause is 'idiotic'. Why? Because four Hiroshima bombs per second, you idiot, that's why. Now look at this poor polar bear and feel the guilt, you ghastly planet-warming human.

People struggling to detect the bullshit in the climate change narrative by now probably aren't cut out for critical thinking.

1.17

The Superbowl of Data Tampering

Temperature readings do matter.

It isn't just bullshit in the narrative we need to be wary of, but the actual data itself. Temperature data is continually revised to make older temperature readings lower and recent readings higher, which has the effect of exaggerating increases or manufacturing increases where they do not even exist. This statement is not made lightly. It is a documentable fact that, time and again, the guardians of the datasets have amended historical temperatures, seemingly always leading to a steeper trend in warming.

Certain people get very uppity when this is pointed out and demand to know if an accusation is being made against NASA, NOAA, the Met Office and others of being in some grand conspiracy. To make that allegation would require evidence of collusion, so let us just stick to the facts we can prove. Here is one such example of temperature alteration described Dr Roger Higgs (DPhil, Oxford, Geology). His July 2021 paper titled *NASA adjusts global temperature graph again* showed how NASA had adjusted a data point in the annual average global temperature for 2016.[118] The before and after readings show a slight lowering, indicated by the downward pointing arrow on the right-hand side.

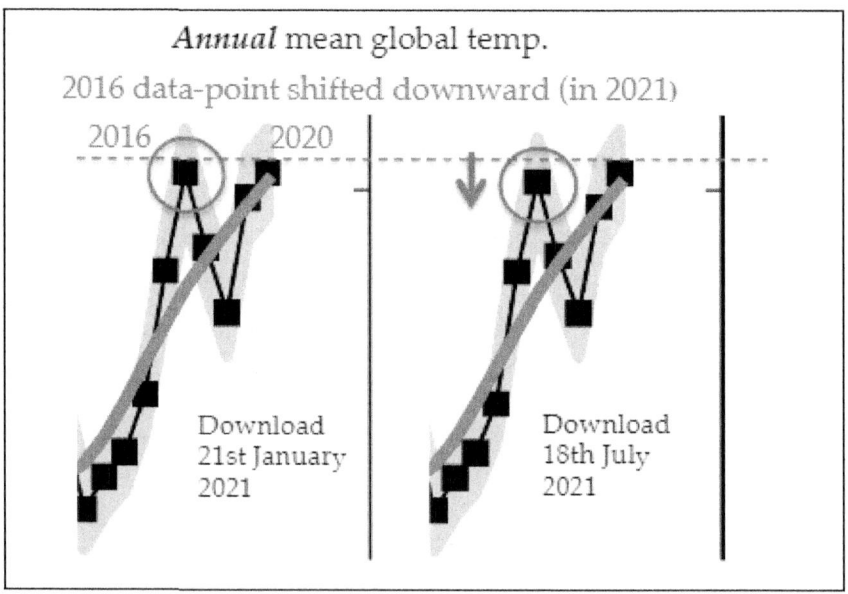

It is a very subtle amendment but it enabled NASA to state that:[119]

"Earth's global average surface temperature in 2020 tied with 2016 for the warmest year on record, according to an analysis by NASA. The year was slightly warmer than 2016 but within the margin of error of the analysis, making the years effectively tied. 'The last seven years have been the warmest seven years on record, typifying the ongoing and dramatic warming trend,' said GISS Director, Gavin Schmidt."

Tony Heller has been exposing these alterations in the temperature record for over a decade now, with numerous essays posted to his website on the subject such as: *Rewriting the climate at NASA*[120], *Understanding NOAA US temperature fraud*[121] and *The Superbowl of data tampering.*[122]

Tony Heller is thoroughly despised by the alarmists for his efforts, not least for the ease with which he dispatches his detractors' criticisms and his great sense of humour. In response to the statement from NASA that:

"Climate-warming trends over the past century are extremely likely due to human activities," Heller noted wryly. "I agree with them – the warming trends over the last century are primarily due to human activities – data tampering by organizations like NASA and NOAA. For example, over the past 20 years, NASA has turned a 70-year cooling trend in the US from 1930 to 2000 into a warming trend."[123]

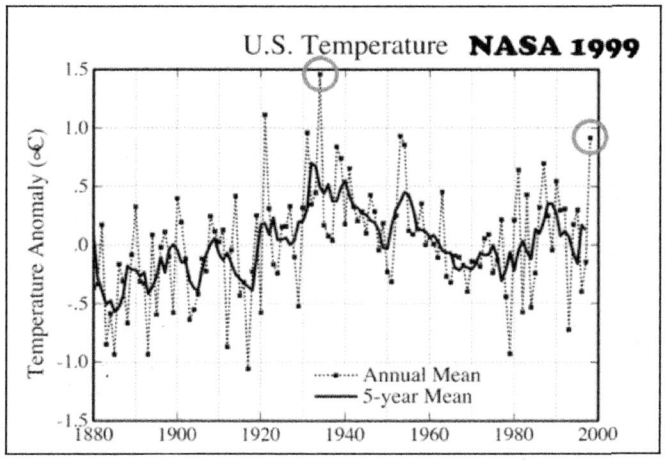

Cooling trend 1940 to 2000

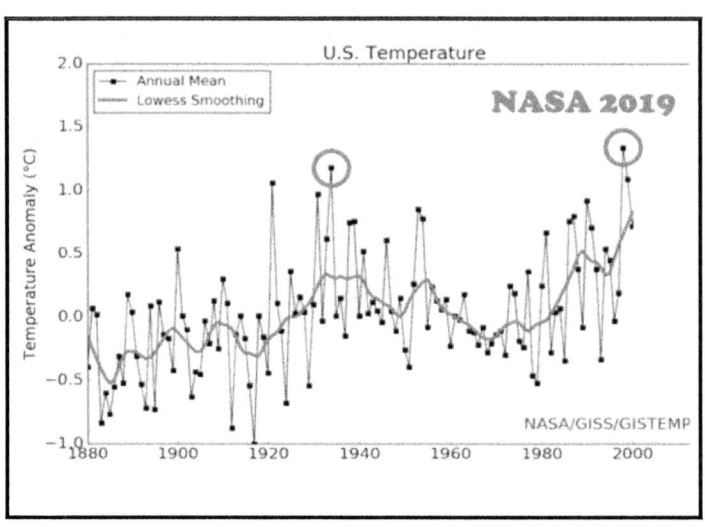

Warming trend 1940 to 2000

If an official temperature reading is taken in 1934 or 1998 and reported in newspapers and scientific journals, one would have thought that such a record should stand. Is it likely that making adjustments in 2019 will give a more accurate result than the one recorded at the time? While there *might* be legitimate scientific reasons for making these adjustments, one would expect adjustments to balance if many of them are made – which they are. When they always end up having the same effect and increasing the warming trend, it raises suspicions.

These adjustments can be made by NASA unilaterally, as has happened with Iceland's temperature record. Tony Heller produced a before and after graph for Iceland's capital Reykjavik.[124]

This is the raw data:

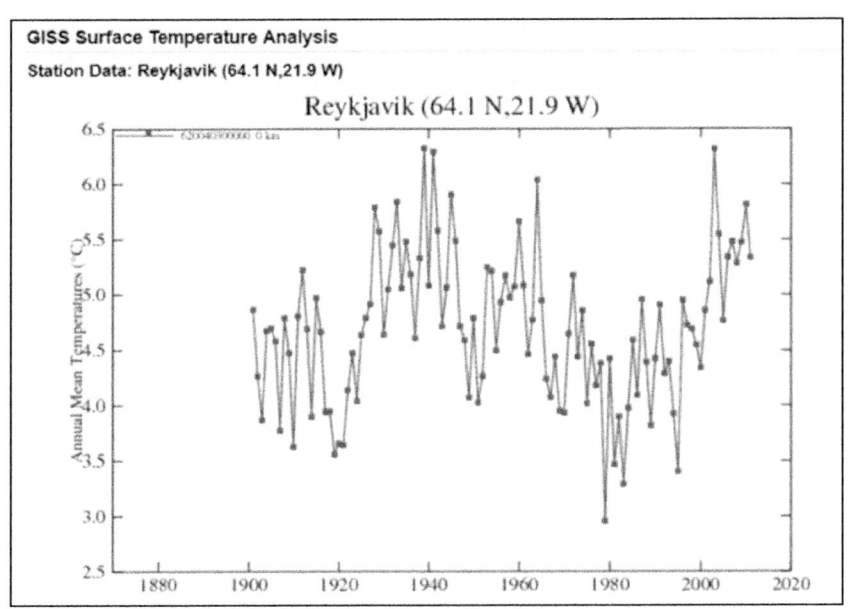

And this is the data after NASA made its adjustments:

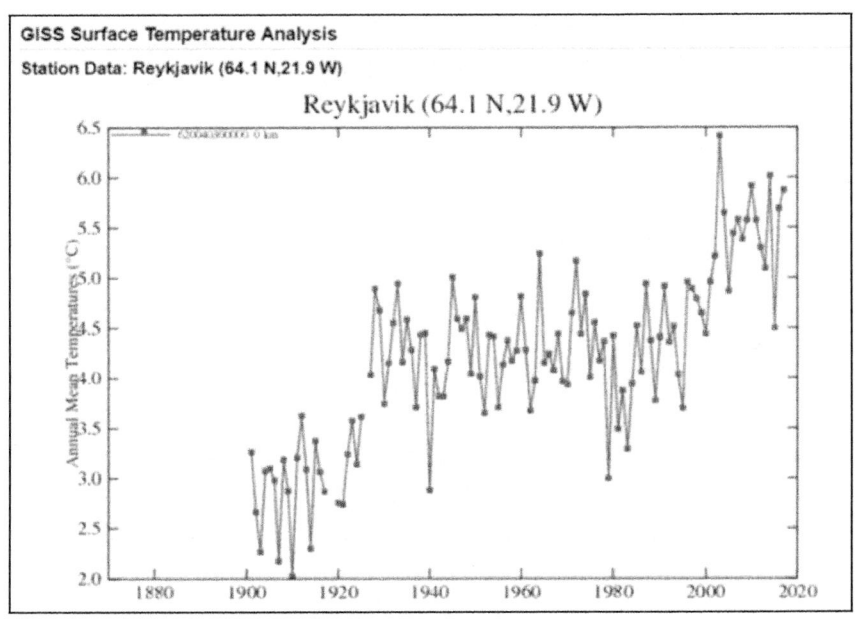

Spot the difference?

Again, a warming trend is manufactured by lowering older temperatures. When a concerned citizen raised this issue with the Icelandic Met Office and asked if the Met Office was aware that its data had been amended, the reply was an emphatic 'no' and an insistence that the original data were just fine:

— Forwarded Message —

From: Trausti Jónsson
To: paul homewood
Cc: Halldór Björnsson
Sent: Monday, 23 January 2012, 17:40
Subject: Re: monthly temperatures

Hi Paul,

We have sent a questions to the GHCN database regarding this and they will look into the problem. Regarding your questions:

a) Were the Iceland Met Office aware that these adjustments are being made?

No we were not aware of this.

b) Has the Met Office been advised of the reasons for them?

No, but we are asking for the reasons

c) Does the Met Office accept that their own temperature data is in error, and that the corrections applied by

GHCN are both valid and of the correct value? If so, why?

The GHCN "corrections" are grossly in error in the case of Reykjavik but not quite as bad for the other stations. But we will have a better look. We do not accept these "corrections".

d) Does the Met Office intend to modify their own temperature records in line with GHCN?

No.

2012 email from the Icelandic Meteorological office.

Strangely, sometime later, Trausti Jonsson reversed his account completely and agreed the adjustments were 'quite sound'.

Removing the warmth of the 1940s without justification was discussed by climate experts in an infamous 2009 Climategate email:

```
From: Tom Wigley <wigley@ucar.edu>
To: Phil Jones <p.jones@uea.ac.uk>
Subject: 1940s
Date: Sun, 27 Sep 2009 23:25:38 -0600
Cc: Ben Santer <santer1@llnl.gov>

<x-flowed>
Phil,

Here are some speculations on correcting SSTs to partly
explain the 1940s warming blip.

If you look at the attached plot you will see that the
land also shows the 1940s blip (as I'm sure you know).

So, if we could reduce the ocean blip by, say, 0.15 degC,
then this would be significant for the global mean -- but
we'd still have to explain the land blip.

I've chosen 0.15 here deliberately. This still leaves an
ocean blip, and i think one needs to have some form of
ocean blip to explain the land blip (via either some common
forcing, or ocean forcing land, or vice versa, or all of
these). When you look at other blips, the land blips are
1.5 to 2 times (roughly) the ocean blips -- higher sensitivity
plus thermal inertia effects. My 0.15 adjustment leaves things
consistent with this, so you can see where I am coming from.

Removing ENSO does not affect this.

It would be good to remove at least part of the 1940s blip,
but we are still left with "why the blip".
```

Climate scientists have been doing this kind of thing for years now. Of course, Michael Mann removed the medieval warming period completely.

Most people who didn't know any better would probably choose NASA's side over Tony Heller's in an argument about temperatures. With NASA being a mightily prestigious organisation, one might be tempted to take NASA's side of the argument and somehow convince oneself that all of those adjustments must have been fair. That might be unwise.

With six degrees, Heller is just as intelligent as most of NASA's staff, if not more so, but most of all he has integrity and devotion to the scientific method.[125] And is NASA really that impressive an authority worth appealing to?

Gavin Schmidt is the Director of NASA's Goddard Institute for Space Studies (GISS), the body responsible for making these temperature adjustments. In 2013, Schmidt agreed to a rather embarrassing television appearance on an American talk show presented by John Stossel, but only on the condition it was not a 'debate'.[126]

In order to cater to Schmidt's whims, the other guest, climate scientist Roy Spencer, had to leave the studio and return after Schmidt had spoken. Unimpressive as it was that NASA would refuse to debate an issue that supposedly affects every living creature on the planet, Schmidt was evasive during the conversation and repeated a lot of the disaster fear porn we hear on mainstream media, much of it based on inadequate computer modelling.

Schmidt claimed unconvincingly that he wasn't prepared to partake in debates 'just to make good TV' and added, "I'm not a politician". Well, Al Gore isn't a climate scientist. Nor do we *want* politicians debating climate change. The Director of NASA's Goddard Institute is the exact person who should be debating the issue. The ironic thing was that during the programme Schmidt opined on what policies we should be enacting to combat the alleged global warming threat, just as if he were a politician.

Schmidt's reticence to debate should not surprise us. Given GISS's track record, one imagines there are a lot of issues that he would have wanted to avoid, added to which Schmidt has no qualifications in climate science, climatology, physics, astronomy, geology or any science subject.[127]

Yes, that's right. NASA's guardian of the temperature data sets is not a climate scientist or a scientist of any description. Gavin Schmidt is a mathematician, a modeller.

As well as observing Schmidt's low-quality performance and thinking twice, one might also ask why NASA hasn't taken legal action against Tony Heller yet. Heller has been accusing NASA of fraudulent behaviour for over a decade. Most importantly, one should ask the simple question: why does NASA keep adjusting the data?

Having blocked Tony Heller on Twitter, this was Schmidt's pathetic response when another Twitter user raised a query on Heller's behalf:

Not the most professional or scientific reply. Nice mask though.

1.18

The Coming Solar Minimum

What if I told you that global warming was not in fact happening at all and that the earth is about to enter an extended cooling period? Would you believe that?

Well why should you? Of course, this author doesn't know. I already said I'm not a scientist and I'm certainly not clairvoyant, but what if NASA told you? What if NOAA told you?

NASA and NOAA are perennial warm mongers but appear to have known for some time that we are entering a solar minimum, even though they do not shout about it.[128] NASA found an interesting way of making the point in a January 2019 article titled:

"Solar Activity Forecast for Next Decade Favorable for Exploration"[129]

Rather than announce a solar minimum as potential global cooling or mention that global warming fears might have been a little overplayed, it was spun as 'good news for mission planners who can schedule human exploration missions during periods of lower radiation, when possible.'

Well, that's one way of looking at it. We might be shivering on planet earth but it'll be good news for space missions! Getting to the crux of the matter, NASA explains that the sun operates on an approximately (but not exactly) 11-year cycle and that cycle strengths can vary.

"The forecast for the next solar cycle says it will be the weakest of the last 200 years. The maximum of this next cycle – measured in terms of sunspot number, a standard measure of solar activity level – could be 30 to 50% lower than the most recent one. The results show that the next cycle will start in 2020 and reach its maximum in 2025."

So, NASA is forecasting a coming cycle that will be 'the weakest of the last 200 years'. This is a huge prediction of the utmost importance and one that you might expect to garner a little bit more attention than it has.

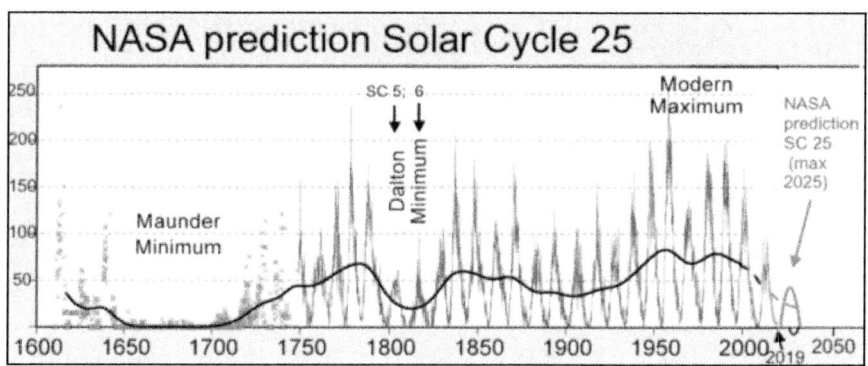

A solar minimum does not mean the end of warm weather and heatwaves, or ice at the equator and frozen waterfalls, but lower solar activity does mean slightly lower temperatures across the globe. Just to be clear, solar cycle 25 has already begun according to a September 2020 press release from NASA titled: *Solar cycle 25 is here.*[130]

The National Oceanographic and Atmospheric Administration (NOAA) has an even stronger view than NASA. NOAA predicts that solar cycle 25 will peak in the year 2025, after which the number of sunspots will fall... and fall... and keep falling into solar cycle 26, and carry on falling all the way until 2040 at least. Throughout the 2030s, NOAA is predicting all but *zero* sunspots. NOAA appears to be predicting a grand solar minimum.[131]

In a grand solar minimum, sunspots can remain at zero for decades and it can get unusually cold. During the Maunder Minimum of 1645-1715, falling temperatures caused famines and the deaths of millions across the world.

Concurrent to the risk of a grand solar minimum, there is another issue to contend with – the Earth's weakening magnetic field. Our magnetic field protects us from harmful cosmic rays and radiation from space. By some estimates, it has lost 30% of its intensity over the last 3,000 years and is still weakening.[132]

With these two wild cards in play, things could get very interesting on planet Earth.

We will only know if these predictions are correct when the time comes. With that said, numerous other institutions and scientists are predicting a solar minimum or even a grand solar minimum. For what it is worth, the 'conventional wisdom' is that yes, there will be a solar minimum and that it has already begun. It is quite extraordinary that at the start of a solar minimum – something history shows will always lead to colder temperatures – we are still exhorted to worry about global warming.

NASA played its part to portray this narrative, downplaying the effects of the solar minimum with an article in February 2020 titled: *There is no impending mini ice age*[133]

This is something of a straw man because the Earth can be subject to a solar minimum without there being a mini ice age. In the article NASA asks: "If such a Grand Solar Minimum occurred, how big of an effect might it have?"

NASA attempts to make the argument that it would not lead to any cooling at all because – you guessed it – manmade climate change trumps all. The overall conclusion in the article is that: "Even a prolonged Grand Solar Minimum or Maunder Minimum would only briefly and minimally offset human-caused warming."

To arrive at this conclusion, NASA made the following argument, asserted without calculations or reference to any academic papers:

"In terms of climate forcing – a factor that could push the climate in a particular direction – solar scientists estimate it would be about -0.1 W/m^2, [Watts per square metre] the same impact of about three years of current carbon dioxide (CO_2) concentration growth.

Thus, a new Grand Solar Minimum would only serve to offset a few years of warming caused by human activities.

What does this mean? "The *warming caused by the greenhouse gas emissions from the human burning of fossil fuels is six times greater than the possible decades-long cooling from a prolonged Grand Solar Minimum.*" (Emphasis added.)

Six times greater than a solar minimum? Considering that we have only had about one single degree of warming over the past century, it is quite some prediction that a solar minimum would only equate to one-sixth of the human effect. And surely there has to be a contradiction with NASA's other prediction that we would see the *weakest* solar cycle in 200 years?

The article becomes even less convincing when one reads the following:

"The current solar cycle, Solar Cycle 24, began in December 2008 and is less active than the previous two. It's expected to end sometime in 2020. *Scientists don't yet know with confidence how strong the next solar cycle may be.*" (Emphasis added.)

This is where things become absurd. How on earth is NASA able to claim 'six times greater' when it admits it does not know how strong the next solar cycle will be in the very same article? The 'six times greater' part is seriously debatable in any case, when one looks at the historical warming using real, not fudged, data.

NASA makes a bizarre appeal to authority – to the IPCC no less – and shockingly refers to the 'scientific consensus'.

"According to the United Nations' Intergovernmental Panel on Climate Change (IPCC), the current scientific consensus is that long- and short-term variations in solar activity play only a very small role in Earth's climate. Warming from increased levels of human-produced greenhouse gases is actually many times stronger than any effects due to recent variations in solar activity."

Are we supposed to be convinced by the fact NASA and the IPCC are on the same page? This is desperate stuff. Grand solar minimums can cause worldwide famines due to freezing but we are supposed to trust NASA's bland assurances that this time they won't because man's fossil fuel burning will have six times the warming effect, but actually we don't know how strong the solar cycle will be even though we said it will be the weakest for 200 years and please trust us, because the IPCC agrees with us.

I don't care how clever they are over at NASA, this is bullshit. It is impossible to predict what NASA is predicting. Impossible. The most honest thing NASA tells us in these articles is that:[129]

"One challenge for researchers working to predict the sun's activities is that scientists don't yet completely understand the inner workings of our star."

Ain't that the truth?

If one is ever asked, "Would you believe something because NASA or NOAA said it?" hopefully we know by now that the source of the claim is irrelevant. Besides, why should we trust NASA or anyone for that matter? For all its brainpower, NASA and NOAA are still government-funded bodies and it's not as if they are above pushing agendas. What matters is whether the things being said are true and what evidence there is for the claim. We should believe in well-reasoned arguments, not a particular source, brand or guru.

Finally, if we do go into a solar minimum, NASA really will have its work cut out pretending that the temperature is still rising. Perhaps now it is easier to understand why the IPCC is so reluctant to study the effects of the sun on climate change.

1.19

Who Can You Trust These Days?

In a world where there was fair reporting, the facts that CO_2 increased in 2020 during lockdown and average global temperatures paused for 20 odd years in spite of significant CO_2 increases, should have been front-page news. After all, it was good news! The pessimists have less to worry about and there is no need to go around terrifying kids that we only have 12 years to save the earth. Even if they are not convinced at this point, they certainly have reason to pause (pardon the pun) and think things over.

Why do the alarmists not look at all the good news and change their tune? The lies we have covered so far still persist and that's not even the half of it. Sadly, where the climate change debate is concerned the truth does not seem to matter. Why not? This is the question we really need to get to the bottom of. For all the useful idiots who have been duped and who repeat mantras without thinking, the big players in this deception are not stupid. They know fine well that what they say is not true. Even when their positions begin to look untenable, the deceit adopts a new guise but always with the same culprit – mankind.

When the thermometers did not oblige, the dire warnings of 'global warming' morphed into dire warnings of 'climate change'. Now climate alarmists have moved to blaming all extreme weather events such as hurricanes, cyclones and floods on manmade climate change, as well as fires of course. One might imagine that death tolls from these disasters were increasing over time but this could not be further from the truth.[134]

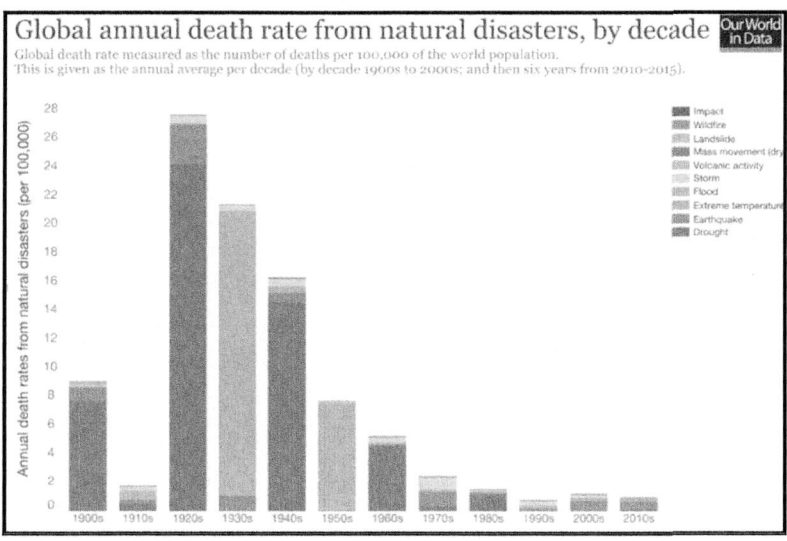

In May 2019, a documentary was aired on the BBC narrated by Sir David Attenborough called 'Climate Change – The Facts', featuring dramatic images of fires, storms and floods which were somehow all man's fault. It was so full of non-sequiturs, one wag joked the title should be changed to 'Climate – *change the facts*'.

Certain people might need to brace themselves for this next part. The media has always given us trusted experts and venerated voices of authority, people the public look up to whom it uses to present information to us. Sir David Attenborough is a classic example, something of a national treasure in Britain and known as the voice of nature after a career spanning many decades presenting wildlife programmes. Most people will hear of no wrong said about a man they hold out as some sort of hero. If you are such a person, well I'm sorry to burst your bubble... but the truth does still matter. Sir David Attenborough has blotted his copybook by boarding the climate change train.

One might excuse him for 'Climate Change – The Facts' and point out that he is just an exceptionally old man reading a script. This is true enough because he is to a large extent a script reader, but what about those walruses?

1.20

Sacred Cows, Fast Horses and Dead Walruses

Sir David Attenborough has helped perpetuate the myth that polar bear numbers have been falling since at least 2011 with his BBC Frozen Planet Series. But in April 2019, he really surpassed himself with the Netflix documentary 'Frozen Worlds', part of the 'Our Planet' series. In this episode, viewers watched a number of walruses falling off cliffs to their deaths on the jagged rocks below, narrated by Sir David Attenborough to sad violin music.[135]

The reason for the carnage in this shocking footage, Attenborough explained, was climate change.

Viewers were told that the walruses only hauled themselves onto land because of a global warming-induced lack of available sea ice –

which Attenborough told us was their natural habitat – before they scaled the cliffs due to a lack of space. By way of explanation for the falling deaths, just before they fell Attenborough said:

"A walrus's eyesight out of water is poor, but they can sense the others down below. As they get hungry they need to return to the sea."

At that point, the footage showed several walruses falling off the cliffs. As the gruesome scenes unfolded in slow motion Attenborough said:

"In their desperation to do so, hundreds fall from heights they should never have scaled."

These are verbatim quotes. Attenborough seems to be suggesting two possibilities: either the walruses could not see properly and accidentally fell or they were in such a rush to eat they took the fastest route to the sea, perhaps forgetting how high they were. As the camera scanned dozens of motionless walrus corpses Attenborough told viewers:

"So many lives of walruses, like those of polar bears and seals, are changing. All are living at the frontier of climate change."

There are so many problems with this account it can only be regarded as deliberate deception.

First of all, the natural habitat of walruses is the oceans where they hunt, not the sea ice as Attenborough claimed. It is possible this was an innocent mistake but he really should know better and so should the scriptwriters. Walruses regularly haul themselves onto land or ice to rest or give birth and they were not hauling out to shore as some kind of emergency due to a lack of ice. In fact, it is known that walruses haul out regularly at this very site.[136]

In addition, it was insinuated that the walruses had climbed the cliff face. This was suggested with the use of close-up footage of a struggling walrus on a steep part of the cliff. Walruses cannot climb cliff faces and they had in fact walked up the slope from the other side.

The location of the incident has been confirmed as Cape Kozhevnikov near the village of Ryrkaipiy in northeast Russia. A contemporaneous news report in the *Siberian Times* sheds a very different light on the event.[137] The article reported that thousands of walruses on the shore attracted another group of big beasts: polar bears. It transpires that a group of these apex predators were responsible for the walruses' deaths. In packs they hunt groups of walruses, charging at them.

Many of the dead walruses at the bottom of the cliffs were likely stampeded to death by other walruses in the panic before others even managed to escape up the slope to the cliffs. Up on the cliffs, they would have been easy to panic and, of course, they had nowhere to run. It is not the first time this has happened nor will it be the last. Naturally, centuries-old incidences of walrus deaths being caused by polar bears cannot be blamed on excess fossil fuel burning.

Although they were not mentioned as a potential contributor to the walruses' deaths, the Netflix team had to be aware that polar bears were in the area because their 'Behind the Scenes' footage blatantly shows them nearby.[138] There is also the sinister possibility that the unnatural sound of the team's flying drone could have frightened the walruses as they huddle tightly in packs and are easily spooked. As the US Fish and Wildlife Service states:[139]

"Walruses often flee haulouts en masse in response to the sight, sound, and especially odours from humans and machines."

It was either a helicopter or a drone that enabled the overhead images to be captured so perhaps some of the deaths were caused by the film crew itself. Either way, it is the height of recklessness to risk scaring these flighty beasts right on the edge of a cliff. One would expect more sense from people who profess to care so much about animal welfare.

In November 2019, a new BBC documentary narrated by Sir David Attenborough called 'Seven Worlds' showed the same walrus deaths in the same location but this time the polar bears *were* shown on film. An annotation by one image on the BBC's website is a clear admission that the polar bears caused the deaths. [140]

"The polar bear's approach causes the walrus to panic and, sometimes, even fall off the cliff."

Screenshot of BBC website, 'Seven Worlds'.

This begs the question: Why did the Netflix team not admit this the first time around? Why did they pretend that climate change caused the deaths and why did they present Sir David Attenborough with an error-strewn script to read?

Even if one is generous or naïve enough to excuse Attenborough from blame, he could have at least issued a statement to clarify his misleading comments in Frozen Worlds.

1.21

I Can't Breathe!

One can forgive Sir David Attenborough for this next deception which was so amateurish it was actually funny. He was the narrator of the documentary 'The Year Earth Changed' and would have had no inkling of the tomfoolery that occurred during the production.

During the programme, viewers were presented with two images side by side, supposedly taken a year apart, purporting to show the difference in air quality following the lockdown of 2020. The message was obvious: pollution is bad and see how lockdowns can totally save the environment guys!

It didn't take viewers long to spot the obvious photo-shopping that was going on as both images had the same two men in the bottom corner wearing identical clothes and standing in the same position and posture, supposedly a year or more apart.

ITV's Lorraine programme featured the image on television in a segment to coincide with Earth Day on 22 April 2021. Her guest, Mike Gunton, who was the producer of the documentary and would or should have been aware of the fakery, said in overly emotional and sentimental tones, "It's almost like Mother Nature saying, 'I can't breathe, I can't breathe!'"

As amusing as this clown show is, we will not be laughing when the crazies in power call for climate lockdowns and, if you think that sounds fanciful, perhaps wait a few years and reassess.

Fun fact: the two men, who are still standing in that exact spot, do not support the Green New Deal or the Great Reset, they voted for Trump and Brexit, do not wear facemasks and they love to burn coal, even in summer, cackling with glee that the Amazon is on fire. They are still dressed in the same clothes.

I'm afraid even the most generous reading of this final incident cannot be attributed to a careless mistake. In the BBC Series 'Blue Planet 2', viewers were shown plastic bags and other pieces of plastic which viewers were informed had been consumed by albatross chicks. There is no footage or actual evidence of the bird consuming this plastic bag. Viewers were just informed that it had happened. Birds do eat plastic and adult birds do feed plastic to their young. This is not a controversial statement. But the claim that an albatross would eat a plastic *bag* is suspicious.

Have you noticed when pigeons are feeding on seeds how they will eat practically whatever amount is put in front of them? Birds have no teeth so they swallow their food whole. Larger birds with claws can rip their food into smaller pieces but most birds cannot do this. What birds do have is a gizzard, where powerful muscles aid digestion by breaking down food. To help this process, birds will swallow hard objects like stones which will make their way into the gizzard. Adult birds will feed small stones to their young and this is all perfectly natural. At sea, a bird might use a smooth piece of plastic for this purpose if it is not able to find a pebble, so if one were to examine the contents of a bird's stomach, it is entirely possible one would find plastic objects.

Birds evolved over millions of years demonstrating survival instincts superior to most species. So, regarding the notion that a bird would just munch on a plastic bag or feed a plastic bag to her chicks, I will believe it when I see it with my own eyes and not on a television documentary from an organisation with a decades' long track record of misleading viewers about climate change.

No one is claiming it is a *good* thing to dump millions of tonnes of plastic in the ocean of course, but this misses the point. A narrative is being sold to an unsuspecting public who do not know any better that their plastics' use is to blame for these birds' deaths when they almost certainly are not. At no point in the episode does Attenborough mention gizzards, or that birds will eat hard objects including plastic quite deliberately and that they will do so all their lives. Are we to believe that Sir David Attenborough is not aware of gizzards and birds' eating habits?

In any case, the biggest source of plastic pollution in the sea by far comes from dumped fishing gear, not shopping bags and straws as we are led to believe.

Again, one might choose to exempt Sir David Attenborough from blame and insist 'he is just reading his lines, he is a very old man, a national treasure and look what he has done in his career, please be nice and why do you hate children and support Donald Trump etc etc.' Angrier readers might ask, who the hell do I think I am – an unqualified nobody criticising NASA and Sir David Attenborough? I understand. We have been taught to practically hero-worship the man and the programming is hard to break.

But I'll say it again: the truth does still matter. As the author Santosh Kalwar said:

"Trust starts with truth and ends with truth."

It is that simple.

Someone once said a man who tells the truth needs a fast horse. Well, if all one is doing is telling the truth, he shouldn't need to run from anyone. Just stand here and watch it sink in. Or not, as the case may be. Fallout be damned if someone can't handle the truth just because he would prefer to believe a lie or cling to false impressions of gurus. This is often the problem. You can show a climate change fanatic a thousand pieces of evidence that contradict his view and he will just double down on his position. As the saying (kind of) goes: 'You can take a horse to water, but you can't make it think.'

Those who would shoot the messenger because they don't like the message are pointing the gun at their own foot. Their problem is not with the messenger but with the truth.

David Bellamy told the truth. He was a wonderful man. He didn't claim he knew better than anyone else how the climate worked; just that there were glaring issues in the story we were being told. Compare the hero worship afforded to Sir David Attenborough to the treatment David Bellamy received for being honest. Like Attenborough, David Bellamy was also a nature and wildlife TV presenter for the BBC, a really enthusiastic educator and all-round lovely man. In 1996, he criticised wind farms on an episode of Blue Peter and after that he found he was no longer welcome at the BBC.[141] After he raised doubts in 2004 about the manmade climate change hypothesis he was let go by the Royal Society of Wildlife Trusts, where he had been President.

Attenborough will probably still be narrating wildlife documentaries when he is 100 and blaming humans for all the supposed ills that have befallen the planet. He is a team player who can be trusted to deliver the 'right' message. Speaking to the UN Security Council – the *Security* Council for goodness sake! – Attenborough said that climate change could destroy entire cities and societies within a lifetime, climate change is the biggest modern threat humans have ever faced and climate change was a threat to global security.[142]

It was pure hyperbole and a transparent appeal to emotion. It was total conjecture on his part and there is no solid evidence to support any of these statements. The fact that it was Attenborough delivering the message does not make it true. When science gives way to emotional blackmail, you can count me out.

At least when Greta rages at the UN while reading scripts prepared by other people, she has the excuse that she doesn't know any better, that she doesn't go to school and has been psychologically damaged by documentaries full of falsehoods about dying polar bears (narrated by you-know-who). Attenborough does not have any of these excuses so his emotionally charged rants – which are really quite pathetic when you take a detached view – suggest he is just acting. Maybe this is what happens when one has worked in television for so long. This is a propaganda onslaught and Sir David Attenborough is part of this machinery. If you don't like that, then I'm afraid that's your tough luck.

John Lennon said, "Being honest may not get you many friends, but it will always get you the right ones." That seems to be the right path to take.

It is fair to say that no human standing on this earth today will ever have a perfect understanding of how the climate works. Spotting the lies in the official narratives is a lot simpler however. The propaganda and lies are endless and the abuse meted out to people who question the narrative is hugely unfair. When I started digging, I found so many problems I ended up asking myself that famous Jeremy Paxman question, "Why are these lying bastards lying to me?"

If an unqualified random like yours truly can spot so many holes in the story how many other lies do you think we are being told? Until such time as our politicians, media and scientific community start displaying a bit more integrity I think a little bit of scepticism is in order. With all this said, it is not the main reason to avoid delving too deeply into the climate change debate. The main reason is that it is not important. It *does not matter.*

Even if the alarmists were 100% accurate in an extreme form that even they probably don't believe – ie that manmade CO_2 is going to cause the sixth mass extinction and the world will end in 12 years...

... or whether it was a complete fairy tale... it does not matter.

It does not matter because it does not matter to the people who are pushing for the Green New Deal. They do not care about climate change. The Green New Deal is about economics, not climate change. The Green New Deal is the gateway to rebooting capitalism. Climate change is the piece of cheese they put on the mousetrap.

I do not imagine for one second that the members of the WEF sit around in Davos worrying about the average world temperature or melting icebergs. They may very well sit around hatching schemes to persuade governments to invest in their friends' carbon capture systems or amend environmental rules. They might well be busy creating plans for smart meters that ration energy use for the rest of us and how to link it to the blockchain. But they aren't worrying about the planet's well-being even if they do pay lip service to it.

Is this not obvious? Rather than listening to what they say, look at what they do. They fly around the world in private jets to conferences to tell people to lower CO_2. They buy beachfront houses that are supposedly going to be swallowed up by rising sea levels. They never ever criticise NATO or the Pentagon, who have the biggest carbon footprints of all. Ever. Just look through their Twitter feeds.

They will happily wind up Greta and the doomsday cultists of Extinction Rebellion like clockwork toys and send them off to spread the message that CO_2 needs to come down to prevent the world temperature rising by that 'fatal' 1.5 degrees. But the chances of the average WEF member believing it in his heart are pretty slim.

No. The Davos crowd is planning bigger things. As we have seen, that thing is the rebooting of capitalism. They have admitted it. They write books about it. You had better believe it.

Some of the least eco-friendly people in the world who are also some of the biggest capitalists in the world, are simultaneously telling *us* that *we* need to be more eco-friendly and also that the world's economic system needs to be reset. They are going to become even richer during this process than they are already. It's actually quite funny when you think about it. Imagine how much contempt they must have for the rest of humanity.

1.22

The Extinction of Sanity

Speaking of colossal amounts of contempt, there is one group of protesters in particular that cannot have escaped your attention. In October 2018, a group called Extinction Rebellion (XR) sprung from nowhere to hit the headlines. They shot to fame shortly after the miraculous rise of Greta Thunberg, like a one-two combo. By the end of 2018, the group had over 100 branches in dozens of countries. Those 'grassroots' activists again.

Extinction Rebellion has three demands which they recite ad nauseam. Vapid propaganda sound bites always seem to come in threes and XR are no different in this regard. The demands are: 'Tell the truth, Act Now, Go Beyond Politics'. The XR website puts a touch more detail on this.[143]

"Tell the truth – Government must tell the truth by declaring a climate and ecological emergency, working with other institutions to communicate the urgency for change.

Act Now – Government must act now to halt biodiversity loss and reduce greenhouse gas emissions to net-zero by 2025.

Go Beyond Politics – Government must create and be led by the decisions of a Citizens' Assembly on climate and ecological justice."

It is fascinating listening to an XR member being interviewed. The member will recite these three demands without fail and, whenever asked a question on anything outside this narrow scope, he or she will tend to clam up or hesitate before returning to the three demands. Members entrusted to speak to the media appear to be under strict orders not to proffer an opinion on anything else at all.

As to the demands themselves, they are ridiculous. Telling the truth 'by declaring a climate emergency' requires one to accept the circular argument that the truth is what XR says it is. This demand is as daft as them declaring the government must tell the truth by announcing that their leader Roger Hallam looks like Brad Pitt.

The demand to reduce greenhouse gas emissions to net-zero by 2025 would be destructive in the extreme because of the importance of CO_2 to every life form on the planet. If XR followed its own advice, to tell the truth, it would realise this.

As to governments being led by a Citizens' Assembly, do we really go to the trouble of electing a government just so they can be 'led by' a Citizens' Assembly that is not elected? We should be extremely sceptical of the claim that a Citizens' Assembly is actually comprised of randomly chosen individuals. XR's utterly transparent desire is to have any such body stacked full of people who 'think the right way' with dissenters from the climate change gospel sidelined. Presumably, these infallible, handpicked individuals would then 'tell the truth'.

Sure enough, in June 2019, the UK government allowed the creation of a Climate Assembly consisting of 108 individuals.[144] One hesitates to say the government 'relented' because one suspects it was a plan that they had signed up to from the start. The UK government informs us that members "were selected from different walks of life, shades of opinion, and from throughout the UK to form a representative sample of the UK's population."[145] For sure, members are representative of the UK's age, geography and ethnicity profile, but it is also claimed that the group's attitudes to climate change were representative of the population.[146]

Criteria	UK population %	Assembly member %	No. assembly members
Very concerned	51.8	49.1	54
Fairly concerned	33.2	32.7	36
Not very concerned	9.3	14.5	16
Not at all concerned	5.1	2.7	3
Other	0.6	0.9	1

Attitudes to climate change (source: Ipsos/Mori, July 2019. Q: How concerned, if at all, are you about climate change, sometimes referred to as 'global warming'?)

Strangely, the group's attitude seems to have changed in a short space of time. As *The Independent* reported in June 2020, before the Assembly's September 2020 final report:

"And 93 percent of assembly members 'strongly agreed' or 'agreed' that, 'As lockdown eases, government, employers and/or others should take steps to encourage lifestyles to change to be more compatible with reaching net zero'."[147]

Quite frankly net-zero is an extremist position and it is astonishing to see 93% approve of it, especially from a group that is claimed to be representative of public opinion. Net-zero will cause harm to them just as it will cause harm to the rest of us and what possible benefits will accrue when countries such as China go full steam ahead building scores of coal-fired electric plants? We all share the same planet.

The Sortition Foundation, which oversaw the selection process, can boast to its heart's content that it has the 'fairest selection algorithm'[148] but something does not make sense. Sortition's fairest selection algorithm seems to have picked a group of zealots happy to self-harm. One wonders exactly what information the 108 Assembly members were exposed to since their joining.

The Climate Assembly was funded partly by the House of Commons and also by two other philanthropic foundations – the Esmée Fairbairn Foundation and the European Climate Foundation.[149] Thus the Citizens Assembly clamoured for by XR and enabled by the UK government is funded by multimillionaire environmentalists. The European Climate Foundation is a private body funded by, among others, Bloomberg Philanthropies, the Climate Works Foundation and the Rockefellers Brothers Fund.[150]

You could not make this up. Yet again the billionaires and philanthropists have tapped their infinite reservoirs of kindness to smooth the wheels toward net carbon zero.

1.23

Putting on a Show

Extinction Rebellion set about making a name for itself by performing one highly publicised stunt after another. In the UK they were stripping off in the House of Commons, climbing up Big Ben, spraying fake blood over the Treasury building and so on. Their tactics were to attract maximum attention – which the media duly gave them – usually by disrupting regular citizens going about their daily business. XR would block London bridges while holding impromptu yoga sessions and glue themselves to office doors or chain themselves to railings until the police removed them.

They wore extravagant costumes and had a selection of props, including an enormous pink boat with the words 'Tell The Truth' written on the side for their April 2019 protest in central London. For the most part, the media loved them. The police treated them with scarcely believable leniency considering the inconvenience and damage they caused. The taxpayer footed the bill for much of it. By

October 2019, the Metropolitan police said it has spent £37 million policing XR's activities.[151] To put that in perspective, this is more than double the £15 million budget of the Violent Crime Taskforce, which was set up to deal with the epidemic of knife crime in London.

Let's not beat around the bush. XR is possibly the fakest, blatantly controlled and well-funded 'opposition' group you are ever likely to see. These props do not come out of a Christmas cracker. Pink boats cost hard cash. And when you see images like this, you have to wonder what is really going on.

How many police does it take to guard a pink boat?

Imagine you gathered a few thousand rebels to protest a worthy cause like child sex abuse and descended on London with all your tents and a great big ornament; say a 20-foot elephant? Would the police allow you to shut down London for the best part of a week and guard your elephant for you? Would the media give such generous coverage to such a protest? Or would they just ignore it as they have been known to do even for protests against NHS cuts that are tens of thousands strong? Anti lockdown protests have seen a *million* people march through London but the media either ignored this completely or claimed: 'thousands' attended, disparaging the protesters as 'conspiracy theorists', 'anti-vaxxers' and 'right-wing extremists' (yawn). Imagine if the media decided to get really stuck into the herberts of XR with the personal attacks? It wouldn't exactly be difficult. Spare a thought for the Yellow Vest protesters in France who are met with extreme violence for having the audacity to peacefully protest about their impossibly high living standards.

Let us do another thought experiment.

Imagine you are the kind of person who watches a lot of BBC news and believes the things that George Monbiot writes in *The Guardian* about climate change. Imagine you really care about CO_2 causing the global average temperature to rise – so you join this groovy group called Extinction Rebellion.

You meet the other members and you have a little 'healing session' with one of their 'climate therapists', then you travel into London with a few friends and a big pot of super glue and glue yourself to a door. Just imagine... and then you chain yourself to the door just in case the glue doesn't stick properly and then your friend disappears taking the key. Now it will require bolt cutters to free you! And then you just wait... and wait. Imagine how you are going to totally save the planet by doing this? Yes, visualise...

(I'm not really going anywhere with this. I just wanted you to try and imagine what goes through the mind of one of these cretins.)

OK, this is not fair to all XR members. Clearly not all of them are cretins. Perhaps we could distinguish between those genuinely good-natured souls of XR who care about the environment and the brainless attention seekers who populate the movement. Many people will have joined XR with honest motives and for many different environmental concerns, all of which are legitimate: deforestation, urbanisation, soil erosion and desertification, industrial farming, the dumping of chemical waste into rivers, over fishing, fracking and so on. Species do indeed go extinct because of corporate greed and no one can sensibly deny this damage.

The problem is that XR mentions virtually none of this. XR's environmental focus is almost exclusively on the one issue of greenhouse gases. Now, if one is going to focus solely on greenhouse gases, let's have a sense of perspective. All humans are not equally to blame for greenhouse gas emissions, despite the guilt we are supposed to feel for flying, driving cars and eating meat. Far from it. You can pin the vast majority of the damage on 100 corporations.

XR is keen to 'tell the truth'. Well here is some truth. A 2017 report showed that just 100 companies are responsible for 71% of global industrial emissions since 1988.[152] Incredibly, a mere 25 corporations were responsible for 51% of global industrial emissions. This is industrial emissions, not those of individuals, and it does not include data about countries' militaries. The Pentagon, for instance, is estimated to have as large a carbon footprint as an entire country like Portugal or Sweden. It is telling how XR never mentions the military-industrial complex and their polluting ways, instead focusing on cars, air travel, meat consumption and plastics use. It always seems to be the fault of the individual.

Remember that even when we include the big industrial emitters of filth and the militaries, human activity only accounts for around 5% of the CO_2 in the atmosphere.

Not only is XR's environmental focus on the one issue of greenhouse gases, but that focus is also nearly all on one gas in particular – CO_2. Just to bang this drum one more time, CO_2 is not a pollutant – something that XR, the media and sadly much of the adult population seems to have forgotten. Vegetable growers add it to their greenhouses to improve yields. Increased CO_2 makes the planet greener. Without CO_2 the world would be a desolate, lifeless wasteland.

Real pollution, on the other hand, consists of things like pesticides, dust particles, ash, heavy metals, CFCs, noxious acids coming from sulphurous and nitrous oxides. Hydrogen sulphide, methane and carbon monoxide are also noxious components of both industrial and natural sources. It is ironic that carbon dioxide is the 'bad guy' when carbon monoxide genuinely is poisonous and exposure to it can kill you in short order.

If you want to talk about pollution, how about shipping? Did you know that just 15 of the world's largest container ships emit as much sulphur oxides as all the world's cars combined?[153] A single huge container ship emits the equivalent of 50 million cars. Shipping is by far the most polluting method of transport in the world. In total, the world's 90,000 vessels emit way in excess of 200 times more sulphur oxides than all the cars combined.

Banning petrol, diesel and hybrid cars will have such a tiny effect on pollution it will amount to almost zero in the scheme of things. When you hear unthinking drones repeat the mantra, 'well it's better than doing nothing', in practical terms it basically isn't. We are being asked to make enormous personal sacrifices for almost zero benefits.

The members of XR with honest intentions have been duped. Just as one has sympathy for Greta Thunberg being used to push an agenda, so one feels sorry for the genuine members of XR who have legitimate environmental concerns. The social engineers have harnessed these concerns and funnelled them all down a blind alley – in this case, the alley labelled 'CO_2'. There is no good evidence that CO_2 causes the temperature to rise, let alone conclusive evidence. CO_2 rose almost uniformly between 1960 and 1990 and has continued to do so right to the present day. Can you see how the steady rise in CO_2 is affecting temperature in the graph below?

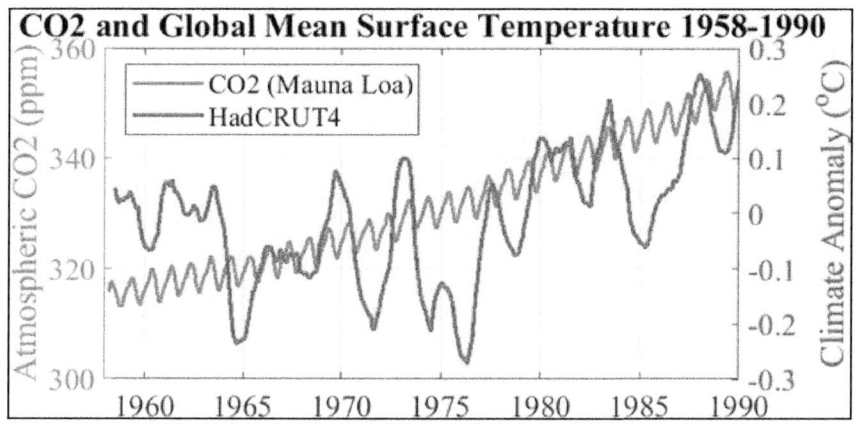

Global average temperature vs atmospheric CO_2 1960 to 1990.

Neither can I.

As we saw earlier, the almost 20-year recent hiatus took place while CO_2 was increasing year on year. This lack of correlation should not surprise us in the slightest because temperatures have never correlated with atmospheric CO_2. We are constantly told that carbon dioxide emissions will lead to temperature increases that will put us all in great jeopardy yet the world has experienced *ice ages* with CO_2 levels ten times higher than today. If CO_2 can be 4,000 parts per million with the temperature around zero, shouldn't CO_2 be even higher today? Why are we supposed to fret about a 1.5 degrees rise in temperature with CO_2 at 417 parts per million?

Proponents of the AGW hypothesis will say, "Ah, but things were different then." The thing is, they always are.

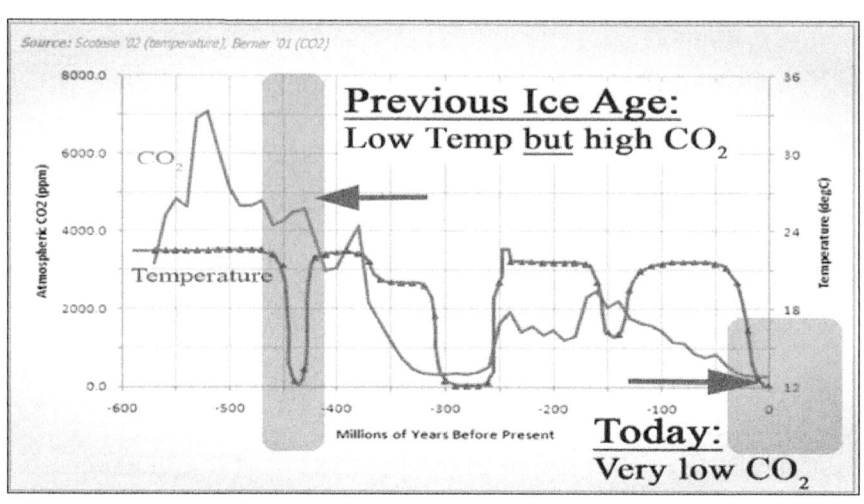

CO₂ at 4,000 parts per million in an ice age? HOW DARE YOU?

I will never claim to understand the workings of climate on planet Earth, but any fool can see that when CO₂ is more than 4,000 parts per million during an ice age there is something wrong with the argument that 20 extra parts per million of manmade CO₂ is going to melt the glaciers and, by the way, can you please stop driving petrol cars and let us tax you into destitution for your own good? The world needs to stop obsessing about CO₂.

Returning to the activities of XR, some of the group's antics were so clownish you had to wonder if they were under the influence of more than just a dim-witted ideology. One morning during the peak London rush hour, a few of XRs foot soldiers mounted the roof of a tube train at Canning Town station, refusing to allow the train on its journey and delaying hundreds of furious commuters.

A word of advice; if you are ever going to try a stunt like this, you aren't going to get a lot of sympathy from the good folk of Canning Town, one of the toughest parts of East London. Indeed, the intrepid environmentalists were dragged off the train and given a portion of shoe pie for their troubles to loud cheers and applause. They could have gotten a lot worse.

If the act of deliberately pissing commuters off as much as humanly possible didn't show how out of touch these people are, their attempt at damage limitation gave us a glimpse into the minds of its leaders. A highly patronising statement was issued which began as follows: "To Londoners, ordinary commuters, working-class citizens, black people, minority communities, humans of all ages. Please read... " Can you think of an introduction more likely to have the readers stop dead in their tracks and finish reading right there?

The amazing thing is that a lot of XR's members are not young people whom one might imagine to be more suggestible. Some of these characters are in their fifties and even older. Most of them have university educations and are quite well to do. The notion that these people are part of a 'rebellion' is hilarious. They are doing the bidding of billionaires who want net carbon zero rules rammed through globally. They are not rebels but useful idiots: storm troopers for the Davos elite.

Some of the histrionics and narcissistic posturing from the group are incredible. When they aren't conducting yoga sessions on Waterloo Bridge, they stage 'die-ins' at Waitrose car park where they all lie down and pretend to be dead. They hold 'confessionals' where they self-flagellate for sins like eating too much meat, using plastic, excess travel and others for ways they have violated the trust of Mother Nature. They set up 'grief tents' at events where fully grown adults congregate to have a good cry about the state of the world while demanding their governments tax them into oblivion. And of course, there are the stunts, often with nudity involved. Something about narcissistic exhibitionism does not tally with the self-loathing and constant talk of the need to make sacrifices. It is as if they are saying, "Humanity is evil and we need punishing, but not us of course. We make sacrifices. We travel to London and throw fake blood over buildings."

One wonders what kind of a reaction they are looking for from their peers and their leaders. Can you imagine the conversation after the tube train stunt?

"Well done, Tarquin, well done, Rupert. Getting your heads kicked in at Canning Town today is going to totally save the world."

"All in a day's work, boss. Next week we will sit in Trafalgar Square and nail our scrotums to the floor. That will definitely bring down the government."

For my next trick...

Speaking of leaders, who is in charge of these people?

Two of the group's figureheads claiming responsibility for this role are Gail Bradbrook and Roger Hallam. Both are hugely unimpressive characters. The odds that either of them conceived of the idea for XR, organised the funding and created a global movement with hundreds of branches, all inside a few months, are a million to one. However, they do hold themselves out as leaders and are the people the media seek out for interviews.

Gail Bradbrook's Wikipedia entry says, "In 2016, she went on a psychedelic retreat to Costa Rica, "where she took ayahuasca, iboga and kambo in search of some clarity in her work." That experience "made her change her approach" to campaigning."

Perhaps Bradbrook transmitted some of this 'clarity' to the group's members. She has also called for mass ingestion of psychedelic substances in protest against the criminalisation of drugs.[154]

Bradbrook compares the XR movement to the struggle of the suffragettes and issues dire warnings about the collapse of civilisation and even the possibility of humanity ceasing to exist altogether in her children's lifetime. It is a wonder that interviewers maintain a straight face at Bradbrook's ravings. However, she did make a rare interesting comment during an interview with Sky News on 18 April 2019. Bradbrook said that the British government was giving XR tacit approval for its protests, claiming:[155]

"The politicians, actually behind the scenes, including this current government, are telling us they need a social movement like ours to give them the social permission to do the necessary."

Adam Boulton (Sky News): "Let's be clear about this, (...) you've actually got government ministers, have you, telling you that's what they want?"

Bradbrook (...) "I've met a couple of people who've talked to Theresa May's, sorry, advisers, and they have said they do know how bad it is and actually they need you guys to help. So, I think basically, we're doing the job (...) we've got three clear demands and the government needs to come and talk to us."

How interesting.

Recall that the UK government's net-zero by 2050 legislation, estimated to cost the UK economy a trillion pounds and probably more, was passed in June 2019 just three months after the April 2019 protests. In the same month, the government did indeed create a Citizens' Assembly for climate change. The government can claim they listened, that they 'told the truth', gave them their Citizens' Assembly, declared a climate emergency and so on. XR gets what they want, which means the people who really control XR get what they want and everyone can pretend that democracy works. It is a false belief: it only works so long as it is acceptable to the vested interests of billionaires and governments themselves. Again, remember the Yellow Vests whose (far more legitimate) protesting earned them a severe beating from the police, not a change in the law and a seat at the table.

If Bradbrook's claim is true, it would certainly explain the police's leniency towards the protesters. There can be no question that Sadiq Khan, the Mayor of London, was content to allow the protests to continue in central London for as long as they did. It would have been a simple matter for him to order the police to clear the area, something they could manage in one night under cover of darkness.

The other figurehead mentioned, could be considered an architect of XR's protests. Roger Hallam is most likely to be heard talking about revolution and bringing down governments. He has the appearance of a man who went to university in the 1970s and never left. Indeed he was studying for a PhD at Kings College London between at least 2017 and 2019, his research question was: *How to achieve social change through civil disobedience and radical movements.*

He has written a document called *Common Sense for the 21st Century*[156] that reads like a 'how-to' guide to carry out protests and guarantee your arrest as efficiently as possible. In this document, he speaks of thousands of people being arrested and sent to prison, claiming that the media publicity would reach millions which would benefit the movement. He states: "It would be beneficial to the Rebellion for people to be in prison before the major civil resistance event to create national publicity" and also that "People can then take time off work, tell their families and prepare for arrest and prison."

Getting arrested and going to prison might just about be acceptable to people who possess the mental and physical fortitude, who have paid their mortgages off and have enough money and no need for a job in the future. But what about literally everyone else? This kind of talk led to concerns that Hallam gave the group a

'Marxist tinge' and whispers started to circulate that Hallam was becoming irate that members were not willing enough to go to prison.

Hallam is actually prepared to practise what he preaches and has been to prison himself, but to seriously expect thousands of your group's members to go to prison doesn't sound much like 'common sense' to me – 'for the 21st Century' or any other.

Roger Hallam ran as an independent in the European Parliament election in 2019, polling 924 votes out of 2,241,681 votes cast in the London constituency. Fun fact: his percentage of 0.04% of the vote is the same as the percentage of carbon dioxide in the atmosphere.

Below is an image of a genuine questionnaire that was discarded during the April 2019 XR protest in London. Among the questions asked was, 'In principle, would you be willing to: Get arrested Y/N? Go to prison Y/N?'

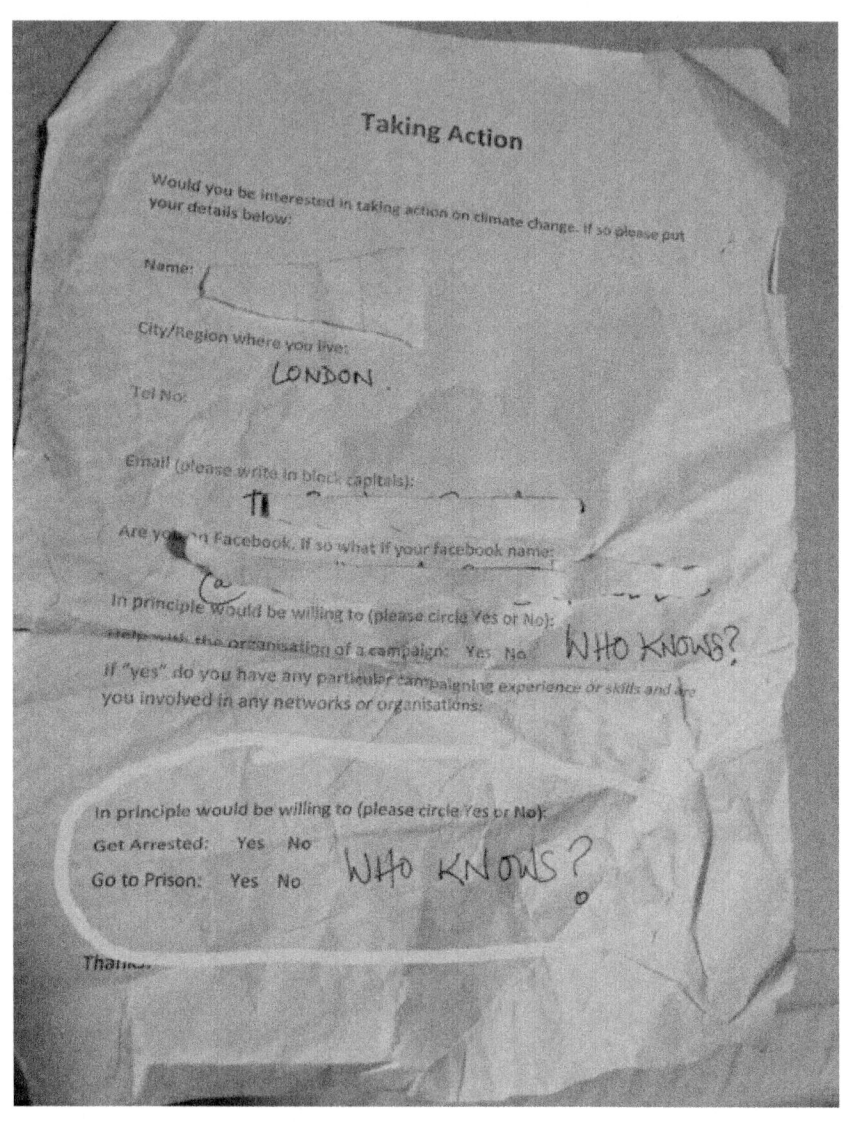

Bear in mind that a lot of youngsters attended these protests with good intentions and had no idea of the people really guiding the movement from behind the curtain. One imagines that the parents of these youngsters would not be too pleased that scruffy, bearded men like Roger Hallam are canvassing their kids to find out which of them are prepared to go to prison. What happens to those who answer 'Yes'? Do they sit and wait for the phone call? Can you imagine such volunteers in the overseas branches of XR where discreet enquiries are made as to whether they might be prepared to start a forest fire to blame on climate change? Not suggesting for a moment that this sort of thing happens, but it bears thinking about it.

One imagines that at the very least those who answer 'Yes' will be logged on some database or other.

Not only that but some are expected to *die*. In July 2019 Hallam said to a crowd:

"We are not just sending out e-mails and asking for donations. We are going to force the governments to act. And if they don't, we will bring them down and create a democracy fit for purpose… and yes, some may die in the process."[157]

While it is not exactly clear how people would die protesting for XR, this kind of talk makes Roger Hallam a dangerous man. One can only imagine what a democracy created by Roger Hallam would look like.

This is where things become sinister. The XR movement has used community centres and even schools to hold meetings. Public sector officials appear willing to accommodate XR, which is concerning. It is bad enough that XR's infantile view of the world and ridiculous warnings of imminent apocalypse and mass death are treated as unchallengeable wisdom by lazy and credulous adults. When the group is recruiting volunteers ready to break the law, to go to prison and even *die* for the cause, they should not be given a platform by local authorities or allowed anywhere near schools.

Another concerning observation is that the group does seem rather obsessed with death. All the talk of mass extinctions, their ridiculous claims that billions of people will die because of climate change, the staging of 'die ins', claiming people might die for the cause, the very emblem they choose for their group (an hourglass symbolising that time is nearly up), all that fake blood and the strange people in red robes conducting what look like rituals. An obsession with death is never healthy.

For a group of supposed do-gooders, XR members have a very misanthropic air about them. It's not just XR though. Hating humanity has become de rigeur among other environmentalist groups with talk of making sacrifices for the good of the planet – including humans – becoming more common.

This is what caused the former president of Greenpeace to leave his organisation in the 1980s. In an interview on 6 March 2019, Patrick Moore said:[158]

"By the mid-1980s, Greenpeace had drifted from the original concept of Green and Peace – meaning environment and people – 'save civilisation from nuclear war'. So, we cared about people and had a humanitarian orientation as well as an environmental one – that was the marriage of green and peace. But, by the time the 80s came along, the peace had kind of got lost and all it was, was 'save the whales' and 'save the earth' and now humans were being characterised as the enemies of nature as if there's too many of us and there shouldn't be so many and people shouldn't have children... I don't believe any of that stuff and at the highest level, I had to get out of Greenpeace at that point. It's way too much like original sin for me, where all the other life is good and only the humans are bad and I just don't buy that. We're from nature just like all the other species are."

It really isn't too far a stretch of the imagination to suppose that one day a mentally enfeebled XR member might take all this green misanthropy to heart and take his own life 'for the cause'. Killing himself to save the earth as it were. Imagine what a tragedy that would be. He would in fact be 'sacrificing' himself for perhaps the fakest, most blatantly controlled opposition movement ever assembled. (That said, Antifa and Black Lives Matter are giving XR a run for its money.)

XR protesters are by and large decent people who want to make a positive difference when it comes to nature and humanity but they are being used as psychological shock troops by people they have not bothered to investigate. The 'climate emergency' is completely bogus. It is hyperbole based on junk science at best and the hysterical ravings of drug-addled imbeciles at worst.

XR has been a controlled opposition outfit from the start. People need to ask themselves what XR's real objective might be before jumping on the climate change bandwagon. The answer is simple: the public needs to believe the bogus climate change crisis narrative in the hope that we will be scared enough to accept draconian new laws in the name of saving the planet. These draconian new laws derive from the Green New Deal and the UN's Sustainable Development Goals. They will be a central part of the WEF's Great Reset.

The environmental movement is dressed in kindness and concern for the earth we inhabit, presented as the Green New Deal with Greta's innocent face and pigtails. It is a phoney narrative. The Green New Deal is actually a giant wrecking ball being taken to the world economy. The WEF and the UN are planning an economic reset to replace capitalism and they have highly placed banking elites on their team ready to carry this out. All of this is openly admitted. How clever to use a sweet little girl simultaneously to drive through brutal change and also as a human shield from criticism.

After the reset, the world will switch over to a technocratic economy based on carbon rationing where everything will be tracked and logged and biometric identities will be implemented to carry out this surveillance. The aim is to switch to a cashless society where as much activity as possible is carried out digitally with virtually all personal information stored on the blockchain. This is not about helping humanity but enslaving it. Make no mistake, the super-rich and the already powerful desire complete control of the world's resources, which includes the population itself, something they regard as just another commodity.

Forget the so-called climate crisis. The real crisis is the authoritarian smart grid being constructed under our noses this very minute, planned by the same billionaires who created the economic bubble which is about to burst.

Playing along with the fake environmental crisis and demanding governments implement austerity measures to 'save the planet', just allows them to create the circumstances they so desperately crave. To join XR is to play right into their hands. Anyone considering joining XR, please, save on the superglue and don't become a weaponised idiot. The best advice to any well-meaning individual wanting to get involved in a real environmental movement is to find a group that hasn't been compromised and co-opted.

How would one know? Apart from being lavishly funded, experiencing instant worldwide growth, attracting saturation media coverage and the police being friendly, here are a few tell-tale signs to suggest a group is not a grassroots movement.

If it is not crystal clear who actually funds the group and it is clearly spending more than it raises, that is a pretty good clue. In fact, if it is well funded at all – especially with billionaire or Silicon Valley donations – that is a red flag in itself. Be suspicious of a group that claims it raised large sums through crowdfunding because you will not be able to audit the source of those funds. Remember that episode of Breaking Bad where Walter White raises money for his son supposedly with an internet appeal? If celebrities and politicians are attaching themselves to the movement, that is another group to avoid. On the other hand, if the government does not approve of an environmental group, that group is probably doing something socially useful.

If its members repeat mantras and do pointless chanting, steer clear. Don't go in for chanting. Groups repeating banal and empty slogans are being mind-controlled, not to mention wasting their time. Vacant exhibitionists who just want to take their clothes off are not usually serious about anything except for likes and shares on social media. If your movement holds healing sessions with climate therapists in the grief tent, it is not a genuine grassroots movement you are dealing with. That is self-indulgent bullshit. And of course, if your group hyper focuses on CO_2 and never, ever mentions NATO, the Pentagon or imperialism, you know you are dealing with controlled opposition.

Better to be knowledgeable in a narrow field, such as soil erosion or pesticide damage, than to be a generalist and a master of nothing.

Specialist knowledge is far more useful than empty sound bites. Read widely. Ignore mainstream media. Switch your television off. It is a minefield and one does have sympathy for the young and the impressionable who really do want to help. Their good intentions can so easily be harnessed and used the wrong way. Just as Greta Thunberg and her follower's good intentions have been hijacked.

One has to wonder why Greta's parents expose a vulnerable child to the international treadmill and media circuit. Who knows what their motives are? You would have to ask them. Perhaps they enjoy the showbiz lifestyle? This line of activity certainly opens up doors to the higher strata of society and the media spotlight. Maybe they love the attention? Perhaps Greta's mother wants to rekindle some interest in her less-than-blockbuster climate change book? Maybe they believe in the cause so passionately they think the risks of burnout or future psychological problems are worth it?

The naughty monkey that sits on one shoulder had a chuckle when I found out that Greta's younger sister Beata Thunberg could be exceptionally unruly and abusive. If the mood took her, she could spend the day screaming profanities at everyone. Yes, I know, one wouldn't wish this on any parent. That said, if anyone deserves being told to 'fuck off' all day long...

One gets the sense that the Greta Thunberg saga is all going to end in tears. The real trajectory of a meteor is not forever upwards and that bright light in the sky comes hurtling down one night at breakneck speed, to land who knows where? The true path of a meteor is downwards. Should that happen to Greta Thunberg, where do you think her handlers will be?

One can hold out hope though. We have seen that she is a headstrong individual and prone to bursts of indignation. Perhaps when she discovers how cynically she has been used she will put this to good use, march up to the UN building and kick their doors down in a fit of righteous anger. As satisfying as that would be, so much will have already been accomplished in Greta's name by then. We will already be firmly on the path of the new economic system the UN and WEF are hell-bent on rolling out.

We should think twice when school kids around the world are targeted to support a long-term economic plan set out for us by billionaires. When the foot soldiers of XR are on the same page as Klaus Schwab, Michael Bloomberg and Christine Lagarde, it ought to give us pause for thought.

The foot soldiers of Black Lives Matter (BLM) also support the destruction of capitalism, just in slightly more aggressive terms. When BLM wants the same thing as the Davos crowd, that should have one's curiosity dials whirring. And who funds BLM? That would be the big corporations, including many WEF members, along with George Soros (quel surprise) and perhaps most interestingly of all – the Ford Foundation – an institution with decades' old links to the CIA. How intriguing.

The elites have been planning this for a long time. XR got roped in at the last minute and handed a few banners to look good for the cameras. They don't have the faintest clue what is really going on. Neither does Greta, bless her.

1.24

The Ghost of Christmas Future

As for the rest of us ordinary people, we are not 'in' on it. In fact, we are very much 'out' of it.

The 99.999% of us are the equivalent of Tiny Tim standing outside Ebenezer Scrooge's house in the cold on Christmas Day, nose pressed up against the window as Scrooge sits down to eat a gigantic turkey.

Which conjures up a vision of Christmas future...

It is almost time for the 2025 annual dust-up at Davos. The great and the good of the WEF are assembled in the finest hotel at the Swiss resort, along with their banker and business leader friends. Several ambitious, beady-eyed politicians are in attendance and even the blood drained vessels of a few faceless UN bureaucrats have been invited. There's a big celebration going on because one of the corporate members won a 20-year contract to supply carbon capture systems to the whole world. No one knows if these systems work yet but the contract was signed anyway.

I approach the window to get a closer look. As I peer in, I can see Klaus Schwab with a glass of champagne in his hand. Michael Bloomberg and Mark Carney are there as well, all smiles.

I shout at him through the window, "I see you Klaus! I know exactly what you are doing."

To my surprise, he turns to face me, looking straight at me somewhat amused.

"And what *are* we doing?" he asks.

"Well, right now you're drinking the finest champagne from a crystal glass that cost more than my laptop. I see you've all still got your yachts and private airplanes while the rest of us are riding bicycles for hire and eating meat grown in a laboratory."

"Very nutritious I've heard. Just wait until the GMO insect protein goes online."

"Can't wait. But I see you, Klaus. I see you smashing up the economy while getting all the big contracts for yourselves. You're taking a wrecking ball to capitalism. That's mafia tactics."

"Not at all, dear boy. We call it 'progress'. The world needs to move with the times. Don't you know, this is all part of the Fourth Industrial Revolution?"

"Yeah, yeah. So where's Greta?"

"Who?"

"Greta Thunberg, remember your Green New Deal girl?"

"Oh yes, Greta." He looks around the room. "Does anyone know what happened to Greta?"

Shrugs all around.

"I'm afraid we don't know what happened to her."

I turn to Mark Carney and Michael Bloomberg.

"Carnage, Bloomberg, I see you. I see you destroying perfectly good businesses. Tell me, why not just smash up the actual polluters if you care about the environment?"

"Sounds like you don't understand what net-zero is old chum," replies Michael Bloomberg. "So long as they are investing in systems to suck that CO_2 back out the atmosphere, they can pollute all they want."

"But no one knows if those systems even work!"

Laughter...

"Hey, do me a favour, Mr Bloomberg. If you are going to handwrite your own rules in order to knee cap businesses to suit your own agenda, can you not call it 'market forces' please?"

Even more laughter.

Mark Carney pipes up, "It's called increased 'transparency'."

Bloomberg and Carney erupt into uncontrollable laughter.

"You don't fool me Carnage, with your 'acceptable face of central banking' image like you're some cuddly vampire squid. We're not falling for this, 'Oh, didn't Mark Carney have a good Brexit' nonsense from your pals in the media. I know what you're doing."

"I can't help it if the media say nice things about me."

"Yeah, like they have a choice! You were Chairman of the Financial Stability Board for seven years while interest rates were floored to zero and the central banks printed 15 trillion. You helped to blow the world's biggest-ever bond market bubble and then legged it, hoping no one would blame the next crash on you."

"Well they didn't, did they?" Carney shoots back with a grin.

"Tell me this though, Carnage. You central bankers can just print money out of thin air and you do, in enormous quantities. So why do you need to unlock all our pensions to invest in your Green New Deal? Why don't you just print the money?"

(Laughs.) "Yes, we thought of that too."

"Well?"

"No comment."

"Why don't you admit that all this environmental scaremongering is bullshit? Admit it."

"Of course it is."

"So you *do* admit it?"

"Listen, if there was a real climate emergency, the politicians and media would all keep quiet about it. That's how the real world works."

"Bloody hell is that sodium pentothal you're drinking?"

"That's why we never raised the alarm about the $300 trillion global debt haystack that was destined to go up flames or the mayhem that would result when the economy rolled over and currencies collapse."

"The debt haystack you helped to create. So what you're saying is if you're in charge, you don't shout fire in a crowded theatre?"

"Exactly. Don't want to alarm people unduly now do we?"

"So what about the Cold War when they scared everyone shitless about nuclear bombs? What about the so-called threats of Russia, or North Korea, or WMDs or any of those? They were quite happy scaring us about that."

Klaus Schwab answers the question.

"Ah, that was different. Scare mongering is only done for *good* reasons. If it's terrorism, your government might need people like you to give up a few freedoms for instance. Do you understand what I mean?"

"Or to sell a few weapons to their mates in the military-industrial complex?"

"Yes, that kind of thing. Or we might need you all to have a vaccine and a biometric identity. For your own good of course and to protect you from a nasty virus, if you get my drift."

"I hear you, Klaus. A virus so bad you needed to reset the world economy."

More laughter.

"So if I've understood you right, Klaus, if the media does have something real to freak out about they never do. And if they are freaking out about something it's probably nothing to worry about."

"Exactly old boy."

"Stuff like the genuine health disasters like obesity, cancer, diabetes and autism rates going off the charts?"

"Yes, we don't want you to worry your little pauper heads about those kinds of things."

"Or fertility rates falling for 60 years and Alzheimer rates skyrocketing?"

"Well, naturally that is good news because we don't want too many of you germ-filled humans running around and reproducing, do we? Now you run along and worry about a harmless gas that only amounts to 400 parts per million."

"Klaus, can I ask you something? You do know we are heading into a grand solar minimum, don't you?"

"Yes, if NASA is to be believed."

"So, not only is your climate scam bullshit, temperatures are going to fall, not rise. Harvests will fail and so on. How are you going to explain this to the world when everyone was worried about global warming yet people are dying because they can't afford their heating bills?"

"We'll tell them that the wonderful green policies they strived so hard for are working! We will say that their sacrifices helped save the earth. The people will be so happy!"

"No, they won't!"

"Yes, they will because we will *tell* them they are happy. The media will run a feel-good campaign to tell them how happy they are. We'll tell them it was all worth it. They'll be as happy as Larry. Never underestimate how stupid you proletarians are, old boy. Besides, by then, who cares?"

"Not you lot I guess. Are we allowed to worry about 5G, Klaus? The media seems very keen not to let us worry about 5G."

"Now listen to me very carefully, old boy. On no account are you ever, ever allowed to criticise 5G. Understand? We can't run a global digital economy without it. *Everything* revolves around 5G."

"I'll expose you. I'll write a book and tell everyone."

"Write away my friend. We care not. You keep saying 'I see you' but it is a glass window you are looking through. We *know* you can see us and that is the whole point. We tell you what we are going to do in our press conferences. We write down our plans in our white papers and we publish them on our websites and in our books."

"You've got a point there, Klaus. Your books are a pile of turgid old shite by the way."

"Perhaps a new ghostwriter is called for. Speaking of ghosts, this conversation only happened in your overactive imagination."

"Yeah. Well, I guess I'll be seeing you, Klaus."

"Not if I see you first."

(Laughter, as the blinds are drawn).

APPENDIX 1 – Bad Science

"No matter if the science for global warming is all phoney. Climate change provides the greatest opportunity to bring about justice and equality in the world."

Christine Stewart – former Canadian Minister for the Environment

"We need to get some broad-based support to capture the public's imagination. So we have to offer up scary scenarios, make simplified, dramatic statements and make little mention of any doubts we might have. Each of us has to decide what the right balance is between being effective and being honest."

Professor Steve Schneider – Stanford Professor of Climatology and lead author of the IPCC reports

"It doesn't matter what is true. It only matters what people believe is true."

Paul Watson – co founder of Greenpeace

"Unless we announce disasters, no one will listen."

John Houghton – first Chairman of the IPCC

"The data does not matter. We are not basing our recommendations on the data. We are basing them on the climate models."

Professor Chris Folland – Hadleigh Centre for Climate Prediction

"This abomination that is occurring today in the climate issue is the biggest threat to the enlightenment that has occurred since Galileo."

Patrick Moore – co-founder of Greenpeace

"The models are convenient fictions that provide something very useful."

Dr David Frame – climate modeller, Oxford University.

"We've got to ride this global warming issue – even if the theory of global warming is wrong. We will be doing the right thing in terms of economic and environmental policy."

Timothy Worth – President UN Foundation

"The threat of environmental crisis will be the international disaster key that unlocks the New World Order."

Mikhail Gorbachev 1996

"Divided nations required common enemies to unite them. Either a real one or else one invented for the purpose. Because of the sudden absence of traditional enemies new enemies must be identified."

"In searching for a new enemy to unite us, we came up with the idea that pollution, the threat of global warming, water shortages, famine and the like would fit the bill. All these dangers are caused by human interventions and it is only through changed attitudes and behaviour that they can be overcome."

"The First Global Revolution – A report by The Club Of Rome" (1991)

"They're brainwashing these children into thinking the world is coming to an end in 10 years. There's got to be billionaires behind this. I want to know who they are. I think we need to hunt them down."

Patrick Moore – co-founder of Greenpeace

APPENDIX 2 – The global hiatus: if you can't explain the pause, you can't explain the cause

"Some people call it a slow-down, some call it a hiatus, some people call it a pause. The global average surface temperature has not increased substantially over the last 10 to 15 years."

Professor Rowan Sutton – Independent – 22 July 2013

"... Despite a sustained production of anthropogenic greenhouse gases, the Earth's mean near-surface temperature paused its rise during the 2000-2010 period... "

Dr Virginie Guemas – Nature Climate Change – 7 April 2013

"We agree with Mr Rose that there has been only a very small amount of warming in the 21st Century. As stated in our response, this is 0.05 degrees Celsius since 1997 equivalent to 0.03 degrees Celsius per decade."

Met Office Blog – Dave Britton - 14 October 2012

"The scientific community would come down on me in no uncertain terms if I said the world had cooled from 1998. OK, it has but it is only 7 years of data and it isn't statistically significant... "

Dr Phil Jones – CRU emails – 5 July 2005

"Bottom line: the 'no upward trend' has to continue for a total of 15 years before we get worried."

Dr Phil Jones – CRU emails – 7 May 2009

"Well, I have my own article on where the heck is global warming... The fact is that we can't account for the lack of warming at the moment and it is a travesty that we can't."

Dr Kevin Trenberth – CRU emails – 12 Oct 2009

"At present, however, the warming is taking a break,"... "there can be no argument about that."

Dr Mojib Latif – Spiegel – 19 November 2009

"It cannot be denied that this is one of the hottest issues in the scientific community... we don't really know why this stagnation is taking place at this point."

Dr Jochem Marotzke – Spiegel – 19 November 2009

"I'm a scientist trying to measure temperature. If I registered that the climate has been cooling I'd say so. But it hasn't until recently – and then barely at all. The trend is a warming trend."

Dr Phil Jones – BBC – 13 February 2010

[Q] "Do you agree that from 1995 to the present there has been no statistically-significant global warming?"

[A] "Yes, but only just."

Dr Phil Jones – BBC – 13 February 2010

"… it has been unclear why global surface temperatures did not rise between 1998 and 2008…"

Dr. Robert K Kaufmann – PNAS – 2 June 2011

"There have been decades, such as 2000-2009, when the observed globally averaged surface-temperature time series shows little increase or even a slightly negative trend (a hiatus period)…"

Dr Gerald A Meehl – Nature Climate Change – 18 September 2011

"The 5-year mean global temperature has been flat for a decade, which we interpret as a combination of natural variability and a slowdown in the growth rate of the net climate forcing.

Dr James Hansen – NASA GISS – 15 January 2013

"… the increase over the last 15 years was just 0.06 degrees Celsius (0.11 degrees Fahrenheit) – a value very close to zero… If things continue as they have been, in five years, at the latest, we will need to acknowledge that something is fundamentally wrong with our climate models… "

Dr Hans von Storch – Spiegel – 20 June 2013

APPENDIX 3 - The IPCC – infallible experts?

Here are 46 enlightening statements by IPCC experts *against* the IPCC:

Dr Richard Courtney: "The empirical evidence strongly indicates that the anthropogenic global warming hypothesis is wrong."

Dr Kiminori Itoh: "There are many factors which cause climate change. Considering only greenhouse gases is nonsense and harmful."

Dr Vincent Gray: "The [IPCC] climate change statement is an orchestrated litany of lies."

Dr Robert Balling: The IPCC notes that "No significant acceleration in the rate of sea-level rise during the 20th century has been detected." This did not appear in the IPCC Summary for Policymakers.

Dr Lucka Bogataj: "Rising levels of airborne carbon dioxide don't cause global temperatures to rise... temperature changed first and some 700 years later a change in aerial content of carbon dioxide followed."

Dr John Christy: "Little known to the public is the fact that most of the scientists involved with the IPCC do not agree that global warming is occurring. Its findings have been consistently misrepresented and/or politicized with each succeeding report."

Dr Rosa Compagnucci: "Humans have only contributed a few tenths of a degree to warming on Earth. Solar activity is a key driver of climate."

Dr Judith Curry: "I'm not going to just spout off and endorse the IPCC because I don't have confidence in the process."

Dr Robert Davis: "Global temperatures have not been changing as state of the art climate models predicted they would. Not a single mention of satellite temperature observations appears in the IPCC Summary for Policymakers."

Dr Willem de Lange: "In 1996, the IPCC listed me as one of approximately 3000 'scientists' who agreed that there was a discernible human influence on climate. I didn't. There is no evidence to support the hypothesis that runaway catastrophic climate change is due to human activities."

Dr Chris de Freitas: "Government decision-makers should have heard by now that the basis for the long-standing claim that carbon dioxide is a major driver of global climate is being questioned; along with it the hitherto assumed need for costly measures to restrict carbon dioxide emissions. If they have not heard, it is because of the din of global warming hysteria that relies on the logical fallacy of 'argument from ignorance' and predictions of computer models."

Dr Oliver Frauenfeld: "Much more progress is necessary regarding our current understanding of climate and our abilities to model it."

Dr Peter Dietze: "Using a flawed eddy diffusion model, the IPCC has grossly underestimated the future oceanic carbon dioxide uptake."

Dr John Everett: "It is time for a reality check. The oceans and coastal zones have been far warmer and colder than is projected in the present scenarios of climate change. I have reviewed the IPCC and more recent scientific literature and believe that there is not a problem with increased acidification, even up to the unlikely levels in the most-used IPCC scenarios."

Dr Eigil Friis-Christensen: "The IPCC refused to consider the sun's effect on the Earth's climate as a topic worthy of investigation. The IPCC conceived its task only as investigating potential human causes of climate change."

Dr Lee Gerhard: "I never fully accepted or denied the anthropogenic global warming concept until the furore started after NASA's James Hansen's wild claims in the late 1980s. I went to the [scientific] literature to study the basis of the claim, starting with first principles. My studies then led me to believe that the claims were false."

Dr Indur Goklany: "Climate change is unlikely to be the world's most important environmental problem of the 21st century. There is no signal in the mortality data to indicate increases in the overall frequencies or severities of extreme weather events, despite large increases in the population at risk."

Dr Mike Hulme: "Claims such as '2500 of the world's leading scientists have reached a consensus that human activities are having a significant influence on the climate' are disingenuous... The actual number of scientists who backed that claim was only a few dozen."

Dr Yuri Izrael: "There is no proven link between human activity and global warming. I think the panic over global warming is totally unjustified. There is no serious threat to the climate."

Dr Steven Japar: "Temperature measurements show that the climate model-predicted mid-troposphere hot zone is non-existent. This is more than sufficient to invalidate global climate models and projections made with them."

Dr Georg Kaser: "This number [of receding glaciers reported by the IPCC] is not just a little bit wrong, it is far out by any order of magnitude... It is so wrong that it is not even worth discussing."

Dr Aynsley Kellow: "I'm not holding my breath for criticism to be taken on board, which underscores a fault in the whole peer-review process for the IPCC: there is no chance of a chapter [of the IPCC report] ever being rejected for publication, no matter how flawed it might be."

Dr Madhav Khandekar: "I have carefully analysed adverse impacts of climate change as projected by the IPCC and have discounted these claims as exaggerated and lacking any supporting evidence."

Dr Hans Labohm: "The alarmist passages in the IPCC Summary for Policymakers have been skewed through an elaborate and sophisticated process of spin-doctoring."

Dr Andrew Lacis: "There is no scientific merit to be found in the Executive Summary. The presentation sounds like something put together by Greenpeace activists and their legal department."

Dr Chris Landsea: "I cannot in good faith continue to contribute to a process that I view as both being motivated by pre-conceived agendas and being scientifically unsound."

Dr Richard Lindzen: "The IPCC process is driven by politics rather than science. It uses summaries to misrepresent what scientists say and exploits public ignorance."

Dr Harry Lins: "Surface temperature changes over the past century have been episodic and modest and there has been no net global warming for over a decade now. The case for alarm regarding climate change is grossly overstated."

Dr Philip Lloyd: "I am doing a detailed assessment of the IPCC reports and the Summaries for Policy Makers, identifying the way in which the Summaries have distorted the science. I have found examples of a summary saying precisely the opposite of what the scientists said."

Dr Martin Manning: "Some government delegates influencing the IPCC Summary for Policymakers misrepresent or contradict the lead authors."

Steven McIntyre: "The many references in the popular media to a 'consensus of thousands of scientists' are both a great exaggeration and also misleading."

Dr Patrick Michaels: "The rates of warming, on multiple time scales, have now invalidated the suite of IPCC climate models. No, the science is not settled."

Dr Nils-Axel Morner: "If you go around the globe, you find no sea level rise anywhere."

Dr Johannes Oerlemans: "The IPCC has become too political. Many scientists have not been able to resist the siren call of fame, research funding and meetings in exotic places that awaits them if they are willing to compromise scientific principles and integrity in support of the man-made global-warming doctrine."

Dr Roger Pielke: "All of my comments were ignored without even a rebuttal. At that point, I concluded that the IPCC Reports were actually intended to be advocacy documents designed to produce particular policy actions, but not a true and honest assessment of the understanding of the climate system."

Dr Paul Reiter: "As far as the science being 'settled', I think that is an obscenity. The fact is the science is being distorted by people who are not scientists."

Dr Murry Salby: "I have an involuntary gag reflex whenever someone says the science is settled. Anyone who thinks the science is settled on this topic is in fantasia."

Dr Tom Segalstad: "The IPCC global warming model is not supported by the scientific data."

Dr Fred Singer: "Isn't it remarkable that the Policymakers Summary of the IPCC report avoids mentioning the satellite data altogether, or even the existence of satellites – probably because the data show a slight cooling over the last 18 years, in direct contradiction of the calculations from climate models?"

Dr Hajo Smit: "There is clear cut solar-climate coupling and a very strong natural variability of climate on all historical time scales. Currently, I hardly believe anymore that there is any relevant relationship between human CO_2 emissions and climate change."

Dr Richard Tol: "The IPCC attracted more people with political rather than academic motives. In AR4, green activists held key positions in the IPCC and they succeeded in excluding or neutralising opposite voices."

Dr Tom Tripp: "There is so much of a natural variability in weather it makes it difficult to come to a scientifically valid conclusion that global warming is manmade."

Dr Gerd-Rainer Weber: "Most of the extremist views about climate change have little or no scientific basis."

Dr David Wojick: "The public is not well served by this constant drumbeat of alarms fed by computer models manipulated by advocates."

Dr Miklos Zagoni: "I am positively convinced that the anthropogenic global warming theory is wrong."

Dr Eduardo Zorita: "Editors, reviewers and authors of alternative studies, analysis, interpretations, even based on the same data we have at our disposal, have been bullied and subtly blackmailed."

Many thanks to "grumpydenier" who first produced this list at his blog in October 2013.[159]

Credentials of the 46 former IPCC scientists are available at the climatism website [160]

ENDNOTES

1 - https://en.wikipedia.org/wiki/Greta_Thunberg#School_strike_for_climate
2 - https://www.wired.co.uk/article/greta-thunberg-climate-crisis
3 - http://www.wrongkindofgreen.org/2019/01/17/the-manufacturing-of-greta-thunberg-for-consent-the-political-economy-of-the-non-profit-industrial-complex/
4 - https://medium.com/wedonthavetime/the-wedonthavetime-manifest-a2f33ce46e63
5 - http://www.wrongkindofgreen.org/2019/10/06/the-global-climate-strikes-no-this-was-not-co-optation-this-was-and-is-pr-a-brief-timeline/
6 - http://web.archive.org/web/20210703153654/https://www.wemeanbusinesscoalition.org/
7 - https://www.cnsnews.com/news/article/barbara-hollingsworth/un-s-top-climate-official-goal-intentionally-transform-economic-0
8 - http://web.archive.org/web/20161125135500/https://www.weforum.org/agenda/2016/11/shopping-i-can-t-really-remember-what-that-is/
9 - http://www.thegwpf.com/ipcc-official-climate-policy-is-redistributing-the-worlds-wealth/
10 - https://news.sky.com/story/court-finds-imf-chief-christine-lagarde-guilty-of-criminal-negligence-10702431
11 - https://www.globalresearch.ca/regime-change-at-the-imf-the-frame-up-of-dominique-strauss-kahn/24866
12 - https://www.ecb.europa.eu/press/key/date/2021/html/ecb.sp210125~f87e826ca5.en.html
13 - https://norberthaering.de/en/power-control/g30-new-investigation/
14 - https://www.bankofengland.co.uk/news/2019/december/mark-carney-named-united-nations-special-envoy-for-climate-action-and-finance
15 - https://www.dailymail.co.uk/news/article-7308313/Companies-face-bankruptcy-fail-prepare-climate-change-Bank-England-boss-warns.html
16 - https://www.theguardian.com/environment/2019/oct/13/firms-ignoring-climate-crisis-bankrupt-mark-carney-bank-england-governor
17 - https://www.blackrock.com/corporate/investor-relations/larry-fink-ceo-letter
18 - https://expertinvestoreurope.com/time-for-mandatory-tcfd-reporting-says-mark-carney/
19 - https://www.theguardian.com/business/2019/oct/08/corporations-told-to-draw-up-climate-rules-or-have-them-imposed
20 - https://www.slideshare.net/CDSB/review-of-tcfd-2019-status-report-slides
21 - https://www.ft.com/content/a775b55a-c5c2-11e9-a8e9-296ca66511c9
22 - https://worldpopulationreview.com/country-rankings/poorest-european-countries
23 - https://www.cnbc.com/2019/08/23/cnbc-exclusive-cnbc-transcript-federal-bank-of-england-governor-mark-carney-speaks-with-cnbcs-steve-liesman-from-fed-summit-in-jackson-hole-wy-today.html
24 - https://www.bankofengland.co.uk/-/media/boe/files/speech/2019/the-growing-challenges-for-monetary-policy-speech-by-mark-
24i - https://news.cityoflondon.gov.uk/financing-roadmap-for-the-net-zero-transition-to-be-launched-by-uk-cop26-advisor-and-un-special-envoy-mark-carney-at-green-horizon-summit/

24ii - http://web.archive.org/web/20210730145224/https://www.brookfield.com/about-us/leadership/mark-carneycarney.pdf?la=en&hash=01A18270247C456901D4043F59D4B79F09B6BFBC
25 - https://www.fiji.gov.fj/Media-Centre/Speeches/English/HON-PM-BAINIMARAMA-AT-THE-OPENING-OF-THE-ONE-PLANE
26 - https://statisticstimes.com/economy/countries-by-gdp.php
27 - http://www.thelocal.fr/20150527/are-top-polluters-funding-paris-climate-summit
28 - https://www.bloomberg.com/news/articles/2020-11-16/creo-syndicate-is-the-secret-club-for-billionaires-who-care-about-climate-change
29 - https://www.forbes.com/sites/larrybell/2013/11/03/blood-and-gore-making-a-killing-on-anti-carbon-investment-hype/
30 - http://climatestate.com/2014/02/06/david-de-rothschild-discusses-cost-of-fixing-climate-change/
31 - https://eelegal.org/wp-content/uploads/2016/12/Rockefeller-Way-Report-Final.pdf
32 - https://newspunch.com/rockefeller-funding-ocasio-cortezs-green-new-deal/
33 - https://www.wrongkindofgreen.org/?s=manufacturing+of+greta+thunberg
34 - https://nymag.com/intelligencer/2019/09/greta-thunberg-climate-change-movement.html
35 - https://www.independent.co.uk/news/long_reads/women/greta-thunberg-climate-change-crisis-strike-austism-misogyny-protest-speech-a9127971.html
36 - https://www.irishtimes.com/life-and-style/people/why-is-greta-thunberg-so-triggering-for-certain-men-1.4002264
37 - https://www.theguardian.com/commentisfree/2019/oct/01/greta-thunbergs-defiance-upsets-the-patriarchy-and-its-wonderful
38 - https://www.ft.com/content/843e1fa4-ea26-11e9-85f4-d00e5018f061
39 - https://www.theguardian.com/commentisfree/2019/sep/30/greta-thunberg-enemies-inaction-climate-crisis
40 - https://www.carbonbrief.org/exclusive-bbc-issues-internal-guidance-on-how-to-report-climate-change
41 - https://www.wsj.com/articles/SB119387567378878423
42 - https://www.nzcpr.com/why-i-am-a-climate-realist/
43 - The Delinquent Teenager Who Was Mistaken for the World's Top Climate Expert (Donna Laframboise)
44 - https://www.epw.senate.gov/public/index.cfm/press-releases-all?ID=f80a6386-802a-23ad-40c8-3c63dc2d02cb&Issue_id=
45 - https://www.ipcc.ch/site/assets/uploads/2018/03/WGI_TAR_full_report.pdf
46 - https://blog.world-mysteries.com/science/35-inconvenient-truths-the-errors-in-al-gores-movie/
47 - https://climateaudit.org/
48 - http://www.lavoisier.com.au/articles/greenhouse-science/climate-change/climategate-emails.pdf
49 - https://www.bccourts.ca/jdb-txt/sc/19/15/2019BCSC1580.htm
50 - https://archive.is/NKvdU
51 - https://www.climatedepot.com/2019/10/01/climatologist-dr-tim-ball-gives-first-interview-after-on-victory-over-michael-mann/
52 - http://news.bbc.co.uk/1/hi/sci/tech/3569604.stm
53 - http://news.bbc.co.uk/1/hi/sci/tech/5109188.stm

54 - http://news.bbc.co.uk/1/hi/sci/tech/7592575.stm
55 - https://www.bbc.co.uk/sounds/play/p08143ny
56 - https://volunteerfirefighters.org.au/the-greens-not-climate-change-are-to-blame-for
57 - https://www.aaas.org/news/michael-e-mann-receives-aaas-public-engagement-science-award
58 - http://web.archive.org/web/20200731204304/https://michaelmann.net/
59 - https://news.psu.edu/story/617687/2020/04/28/research/mann-elected-national-academy-sciences
60 - https://www.gov.uk/government/news/uk-becomes-first-major-economy-to-pass-net-zero-emissions-law
61 - https://www.instituteforgovernment.org.uk/explainers/net-zero-target#references
62 - https://www.newsweek.com/leonardo-dicaprio-helps-raise-5-million-amazon-fires-celebrities-call-action-1456200
63 - https://twitter.com/antonioguterres/status/1164586391629705216
64 - https://twitter.com/EmmanuelMacron/status/1164617008962527232
65 - https://twitter.com/KamalaHarris/status/1165070218009489408
66 - https://twitter.com/GlobalEcoGuy/status/1164732075301146624
67 - http://www.yadvindermalhi.org/blog/does-the-amazon-provide-20-of-our-oxygen
68 - https://www.nationalgeographic.co.uk/environment-and-conservation/2019/08/why-amazon-doesnt-really-produce-20-worlds-oxygen
69 - https://www.bbc.com/future/article/20120905-save-our-dying-coral-reefs
70 https://www.nationalgeographic.com/science/article/scientists-work-to-save-coral-reefs-climate-change-marine-parks
71 - https://www.bbc.co.uk/programmes/p04771yp
72 - http://www.bbc.com/earth/story/20140916-the-corals-that-come-back-from-the-dead
73 - https://www.cbsnews.com/news/climate-change-threatens-maldives/
74 - https://www.theguardian.com/global-development/2019/may/16/one-day-disappear-tuvalu-sinking-islands-rising-seas-climate-change
75 - https://www.theguardian.com/environment/gallery/2014/may/30/kiribati-line-in-sand-pictures
76 - https://www.nature.com/articles/s41467-018-02954-1
77 - https://www.nationalgeographic.com/science/article/150213-tuvalu-sopoaga-kench-kiribati-maldives-cyclone-marshall-islands
78 - https://www.resilience.org/stories/2018-02-22/islands-not-sinking-climate-change-demonstrated-to-be-a-hoax/
79 - https://www.nasa.gov/specials/sea-level-rise-2020/
80 - https://sealevel.nasa.gov/faq/13/how-long-have-sea-levels-been-rising-how-does-recent-sea-level-rise-compare-to-that-over-the-previous/
81 - https://royalsociety.org/topics-policy/projects/climate-change-evidence-causes/question-14/
82 - https://www.theguardian.com/environment/commentisfree/2021/apr/13/sea-level-rise-climate-emergency-harold-wanless
83 - https://tidesandcurrents.noaa.gov/publications/techrpt83_Global_and_Regional_SLR_Scenarios_for_the_US_final.pdf
84 - https://www.ipcc.ch/site/assets/uploads/2018/05/ar4_wg1_full_report-1.pdf (p69-70)

85 - https://www.forbes.com/sites/alexepstein/2015/01/06/97-of-climate-scientists-agree-is-100-wrong/
86 - https://skepticalscience.com/97-percent-consensus-cook-et-al-2013.html
87 - https://iopscience.iop.org/article/10.1088/1748-9326/8/2/024024
88 - https://wattsupwiththat.com/2013/05/21/cooks-97-consensus-study-falsely-classifies-scientists-papers-according-to-the-scientists-that-published-them/
89 - https://www.econlib.org/archives/2014/03/16_not_97_agree.html
90 - https://www.climatedepot.com/2015/09/03/its-all-wrong-un-convening-lead-author-dr-richard-tol-slams-media-for-false-claims-about-alleged-97-consensus/
91 - http://web.archive.org/web/20150906020454/https://science.house.gov/legislation/hearings/full-committee-hearing-examining-un-intergovernmental-panel-climate-change
92 - https://climatechangedispatch.com/97-articles-refuting-the-97-consensus/
93 - www.skepticalscience.com/97-percent-consensus-robust.htm
94 - https://www.theguardian.com/environment/2019/nov/05/climate-crisis-11000-scientists-warn-of-untold-suffering
95 - https://twitter.com/MRobertsQLD/status/1191964557608140801
96 - https://twitter.com/righttoclimb/status/1191967198828531713/photo/1
97 - http://www.petitionproject.org/index.php
98 - https://www.epw.senate.gov/public/index.cfm/press-releases-all?ID=f80a6386-802a-23ad-40c8-3c63dc2d02cb&Issue_id=
99 - https://www.bbc.co.uk/news/av/world-51193460
100 - https://www.thelancet.com/journals/lancet/article/PIIS0140-6736(14)62114-0/fulltext
101 - https://www.sciencedirect.com/science/article/abs/pii/S0921818116304787
102 - https://www.che-project.eu/news/how-do-human-co2-emissions-compare-natural-co2-emissions
103 - https://science.nasa.gov/earth-science/oceanography/ocean-earth-system/ocean-carbon-cycle
104 - https://en.wikipedia.org/wiki/Atmosphere_of_Earth
105 - https://research.noaa.gov/article/ArtMID/587/ArticleID/2636/Rise-of-carbon-dioxide-unabated
106 - https://ib.bioninja.com.au/standard-level/topic-4-ecology/44-climate-change/greenhouse-gases.html
107 - https://css.umich.edu/factsheets/greenhouse-gases-factsheet
108 - https://www.geocraft.com/WVFossils/greenhouse_data.html
109 - https://hockeyschtick.blogspot.com/2014/11/updated-list-of-64-excuses-for-18-26.html?
110 - https://www.bbc.co.uk/news/science-environment-24173504
111 - https://www.theguardian.com/environment/damian-carrington-blog/2013/sep/27/global-warming-pause-mirage-ipcc
112 - https://www.westernjournal.com/nat-geographic-admits-wrong-famous-climate-change-polar-bear-pic/
113 - https://www.climatedepot.com/2019/03/05/study-polar-bear-numbers-reach-new-highs-population-increases-to-the-highest-levels-in-decades/
114 - https://polarbearscience.com/
115 - https://skeptics.stackexchange.com/questions/40292/does-susan-crockford-have-any-scientific-credentials-related-to-polar-bears

116 - https://en.wikipedia.org/wiki/Susan_J._Crockford
117 - https://climatechangedispatch.com/lynching-polar-bear-expert-crockford/
118 - https://www.researchgate.net/publication/353339542_NASA_adjusts_global_temperature_graph_again
119 - https://earthobservatory.nasa.gov/images/147794/2020-tied-for-warmest-year-on-record
120 - https://realclimatescience.com/rewriting-the-climate-at-nasa/
121 - https://realclimatescience.com/understanding-noaa-us-temperature-fraud/
122 - https://realclimatescience.com/2020/02/the-superbowl-of-data-tampering/
123 - https://realclimatescience.com/are-government-temperature-graphs-credible/
124 - https://realclimatescience.com/2020/01/nasa-confirms-their-own-conspiracy-theory/
125 - https://realclimatescience.com/who-is-tony-heller/
126 - https://www.drroyspencer.com/2013/04/stossel-show-video-schmidt-vs-spencer/
127 - https://science.gsfc.nasa.gov/sed/bio/gavin.a.schmidt
128 - https://science.nasa.gov/science-news/news-articles/solar-minimum-is-coming
129 - https://www.nasa.gov/feature/ames/solar-activity-forecast-for-next-decade-favorable-for-exploration
130 - https://www.nasa.gov/press-release/solar-cycle-25-is-here-nasa-noaa-scientists-explain-what-that-means
131 - https://www.swpc.noaa.gov/products/predicted-sunspot-number-and-radio-flux
132 - https://riskfrontiers.com/insights/risks-of-weakening-of-the-earths-magnetic-field/
133 - https://climate.nasa.gov/blog/2953/there-is-no-impending-mini-ice-age/
134 - https://ourworldindata.org/natural-disasters
135 - https://www.youtube.com/watch?v=cTQ3Ko9ZKg8&ab_channel=Netflix
136 - https://www.thegwpf.com/gwpf-calls-for-david-attenborough-to-come-clean-on-walrus-tragedy-porn/
137 - http://siberiantimes.com/ecology/others/news/village-besieged-by-polar-bears-as-hundreds-of-terrorised-walruses-fall-38-metres-to-their-deaths/
138 - https://www.ourplanet.com/en/video/behind-the-scenes-walrus/
139 - https://www.fws.gov/alaska/fisheries/mmm/walrus/wmain.htm
140 - https://www.bbc.co.uk/programmes/articles/4zh2Dd3JC8gprNZcGY6BbHB/walrus-on-the-edge
141 - https://www.dailymail.co.uk/news/article-2266188/David-Bellamy-The-BBC-froze-I-dont-believe-global-warming.html
142 - https://thehill.com/homenews/news/540058-david-attenborough-warns-un-security-council-on-climate-change-i-dont-envy-you
143 - https://extinctionrebellion.uk/the-truth/demands/
144 - http://web.archive.org/web/20210812091526/https://www.parliament.uk/get-involved/committees/climate-assembly-uk/
145 - https://www.climateassembly.uk/about/index.html
146 - https://www.climateassembly.uk/detail/recruitment/index.html
147 - https://www.independent.co.uk/news/uk/politics/coronavirus-economic-stimulus-economy-net-zero-carbon-a9580086.html
148 - https://www.sortitionfoundation.org/its_official_we_use_the_fairest_selection_algorithm
149 - https://www.climateassembly.uk/detail/budget/index.html

150 - https://europeanclimate.org/funding-grantmaking/
151- https://www.theguardian.com/uk-news/2019/oct/22/extinction-rebellion-protests-cost-met-police-37m-so-far
152- https://b8f65cb373b1b7b15feb-c70d8ead6ced550b4d987d7c03fcdd1d.ssl.cf3.rackcdn.com/cms/reports/documents/000/002/327/original/Carbon-Majors-Report-2017.pdf?1499691240
153 - https://newatlas.com/shipping-pollution/11526/
154 - https://www.newscientist.com/article/2213787-extinction-rebellion-founder-calls-for-mass-psychedelic-disobedience/#
155 - https://www.youtube.com/watch?v=zvFW6UWKNPM
156 - https://www.rogerhallam.com/wp-content/uploads/2019/08/Common-Sense-for-the-21st-Century_by-Roger-Hallam-Download-version.pdf
157 - https://www.dailymail.co.uk/news/article-7267533/Extinction-Rebellion-taking-Marxist-tinge-say-disillusioned-members.html
158 - https://soundcloud.com/breitbart/breitbart-news-tonight-patrick-moore-march-6-2019
159 - https://grumpydenier.wordpress.com/2013/10/08/46-statements-by-ipcc-experts-against-the-ipcc/
160- https://climatism.wordpress.com/2020/03/07/46-statements-by-ipcc-experts-against-the-ipcc/

PART TWO

2.1

The Big Lie

"If you tell a lie big enough and keep repeating it, people will eventually come to believe it."
Joseph Goebbels

It is difficult to point to one popular myth or hoax and declare 'this is the biggest lie of all'. With that said, if prizes were handed out for popular myths that aren't true, then this particular lie deserves a place near the top of any shortlist. This myth must be refuted because so many other popular myths are constructed on the back of it.

This belief is so ingrained that most people reading this sentence probably believe it themselves, so be prepared to have that belief system challenged. At least be prepared to hear some of the counter-arguments and question whether what we have been told all our lives is true. Let us get this one out of the way early. Here it is, in the author's opinion, *the biggest lie of all*:

That the world is overpopulated.

Contrary to what we are led to believe, the world is not overpopulated by ghastly, polluting, germ-riddled, planet-destroying humans. It is not true, nor is the companion myth that overpopulation will lead to environmental and ecological collapse or a mass extinction event. These things may happen one day but, if they ever do, then it will not be due to too many people on earth.

The claim that the world is overpopulated is a centuries-old lie that we have been conditioned to believe since childhood. It is a lie propagated by the media, academia, think tanks, talking heads and it is the theme of books and Hollywood films. Moreover, it is a lie that is being spread with an increasing sense of urgency, foreboding and more than a hint of misanthropy. It is not uncommon to read dire warnings such as the one made by TV presenter, Sir David Attenborough, in 2013 when he referred to humans as a 'plague on Earth'.[1]

"We are a plague on the Earth. It's coming home to roost over the next 50 years or so...

It's not just climate change; it's sheer space, places to grow food for this enormous horde. Either we limit our population growth or the natural world will do it for us, and the natural world is doing it for us right now."

David Attenborough was parroting the very same misplaced rhetoric as Thomas Malthus from the 18th Century. Malthus' words were not true in 1798 and neither were David Attenborough's words in 2013, two centuries and six billion extra people later. When one deconstructs this myth for the patent falsehood that it is, an obvious question arises: "Why have we been constantly lied to?" Talk of humanity being a 'plague' or a 'virus' or a 'cancer' is not only a lie but a sinister lie.

As the joke goes, a traveller who is hopelessly lost stops a passer-by and asks him if he knows the way to his destination. On hearing the name of the destination the passer-by replies:

"Well, I wouldn't start from here."

If one believes the world is overpopulated, that person is starting from the wrong place. So many other myths are born from the overpopulation myth it is something of a grandfather of myths. We need to shatter the illusion that the world is overpopulated and we need to do so right away. Truth will set you free and all that.

Without delving into the question *why* this myth persists for now, let's get down to the business of refuting it. A wise man once said, "It's not what you don't know that's the problem. It's what you know for sure that ain't so."

Never a truer word spoken.

2.2

We Easily Have Enough Space

City dwellers are used to queues, crowds, packed trains and walking behind slow people, added to which they are constantly told the world is overcrowded. Most of them probably believe it. Put that person on a thirty-hour bus journey between cities in a country like Brazil and he will probably feel differently.

Cities are overcrowded. The world is not.

As of July 2020, an estimated 56.2% of the world's population lived in cities, leaving the entirety of the rest of the Earth's land mass for the remaining rural 43.8%.[2]

This is an enormous amount of space for the rural population because cities only occupy a tiny proportion of the world's area. It is difficult to establish exactly what this proportion is because definitions of 'city' or 'urban area' vary, but estimates vary from 0.3% to 2.7%.[3]

One particular study from 2016, showed that half the planet's population lived on just 1% of the land. Real estate data expert, Max Galka, used NASA's gridded population data to create maps highlighting the population centres. Look at his maps and one realises that the vast majority of the planet is basically empty.[4]

To take an extreme example, the entire world's population could in theory all stand on the Isle of Wight, a small island in the English Channel. For an even more extreme example, this is what the entire world's population might look like piled into the Grand Canyon.

When someone shouts, "Bundle!" and the whole world joins in.

Imagine being at the bottom of that 'bundle'. Granted this is a silly example but if this was real it would mean the rest of the entire world would not have a single person occupying it. How many people do you think could fit on the world's surface and still have adequate living space?

Such calculations have been performed numerous times. One example shows that all the people on earth could comfortably fit into an area the size of Texas and have 93 square metres each. A family of four would have 372 square metres, enough for a good sized house and a garden.

Refuting the claim that the world is overpopulated is not the same as claiming there are no overpopulated areas. Some places are overpopulated because we have been herded into cities for generations, in fact, centuries, where traditionally it was easier to tax the human cattle and, in future generations, it will be easier to surveil us, control us and scrape every last piece of usable data out of us. This trend is set to continue and the UN estimates that 68% of the world's population will be living in cities by the year 2050.[5] The trend might feel organic, but rest assured, it is by design.

2.3

We Have Plenty of Food

As you can see, the world is not lacking space in which to grow food. The world is not lacking food either, something that proponents of the overpopulation myth like to claim.

Of course, not all land is habitable or fertile. One cannot grow wheat or vegetables in the desert or up a frozen mountain but, even so, the world is not lacking space to grow food. Much of the world's land is unused. This is often by design as well. Millions of acres of perfectly good farmland are lying idle because the US government pays farmers not to grow crops.

Just on the farmland that was being used in 2012, enough food was produced to feed every person on the planet plus another three billion people.[6] Looking at the surplus grain mountains and empty farmland, it should be obvious that the problem is not a lack of food or a lack of room on which to grow the food or an ability to produce the food. It is a question of distribution.

Feeding ten billion people is achievable without even increasing the acreage used. Nor does this take into account the inevitable technological advancements that will come our way. (We are not talking about genetically modified crops here either. Not a single mouthful of genetically modified bathed-in-glyphosate produce needs to be consumed.) There are vast untapped physical and human resources available to the world, not to mention the intangible benefits of human ingenuity. We could feed even more than ten billion people and do so very easily. So, why does starvation persist?

Think back to the space race of the 1960s. That project certainly had the attention of the world's superpowers. Imagine the world focussing its resources as it did back then but intending to prevent starvation. All hands to the pump, government budgets diverted in the right areas, incentives aligned and massive publicity campaigns. Just suppose the world's governments attacked starvation with the alacrity they approached the space race or the way they prioritised the vaccine rollouts in the early 2020s. All that effort, but applied to food distribution.

It isn't for a lack of money. The monetary cost of ending hunger worldwide has been estimated at a mere $30 billion per year.[7] Even if $30 billion per year is a lowball estimate we could treble it, then round it up to $100 billion and it would still be less than one-seventh of the annual US 'defence' budget. World GDP is around $80 trillion. We are talking about a number that is around 1/800th of what the world produces each year. $100 billion is *nothing* in the scheme of things.

Reducing starvation is a question of willpower and the will simply does not exist.

As an aside: we need to remember this sentiment if the world does suffer food shortages in the coming years because such a scenario is looking increasingly likely. Look at what is happening around us.

The US government recently expanded a programme that pays farmers to leave land fallow, ostensibly to reduce CO_2 emissions.[8]

This policy will doubtless cause problems because of the simultaneous raft of measures introduced in response to COVID-19 that disrupted supply chains. To be sure, COVID-19 did not cause these supply chain disruptions, the response measures did. While money can be created out of thin air, food cannot, so when supply chains are restricted at the same time as endless money printing funds deficits and stimulus payments, the result is going to be inflation. Food price inflation will cause people at the margins to starve.

COVID-19 restrictions shackled food producers and the measures always seemed to hit meat producers hardest. It appears the war on CO_2 is being expanded to a war on methane, with cows especially in the firing line. Just wait for Bill Gates and Jeff Bezos to ride to our 'rescue' with disgusting, unhealthy, fake meat produced in a laboratory.

Then there is California. California is facing acute water shortages that will have severe effects on food production. California leads all other states in farm income and produces a third of the USA's vegetables and two-thirds of its fruit and nuts.[9] The state's farmers desperately need water.

That California should be facing a water shortage is perplexing because its reservoirs were full again by 2019 after a 7-year drought that began in 2011. The reason is mind-boggling. The reservoirs are being drained into the ocean to protect rare species of salmon. You really couldn't make this up.[10]

Leaving land fallow to reduce CO_2 emissions is ironic considering CO_2 is plant food. All this measure will achieve is food price inflation when shortages kick in. As to the water shortages in California, if the people in charge there genuinely value the lives of a few salmon over the country's food supply, then they are suffering from a form of green insanity. Madness is the only logical explanation unless of course the motive is sabotage. It genuinely does look like a drought is being manufactured.[11]

To a suspicious observer, this looks like a multi-pronged attack on the food supply, always under the excuse of environmentalism and sustainable development. One stupid decision can be explained away but when you see them made as frequently as this, it becomes harder to pass them off as mere accidents.

The world can easily feed ten billion people. Let us not forget that. This is a statement made with the highest confidence but the author is fully aware of how hollow this claim might look in future when people are facing starvation. So remember this; if the world fails to produce enough food for even five billion people you can bet it was planned that way.

* * *

Natural disasters and wars will always lead to some degree of hunger in the world but the problem is usually more to do with politics. Corrupt governments of poor countries are often propped up with foreign aid from corrupt governments of rich countries. Too often conditions and incentives are attached to aid packages that govern population control rather than raising living standards. Corrupt governments have no incentive to raise living standards if foreign aid keeps arriving. Too often those in charge work in the opposite direction to the one that would solve their human problems. There is under-investment in sanitation, contaminated water supplies are not cleaned and stolen land is not returned to the people so that they can grow food and achieve nutritional independence. The poor people of the recipient countries see almost none of the foreign aid that comes in and the last thing their governments are likely to do is to let the people have power. If that happened, they might kick out the local and global corporations that are leeching their natural resources. African leaders who have tried to do right by their people in the past have often found themselves replaced in coups, perishing in mysterious plane crashes, or in Gadaffi's case, sodomised with a bayonet by NATO-backed jihadists. What these countries need is foreign *investment* without strings attached.

If you pay a visit to a website belonging to the UN, the World Bank, the IMF or other such supranational bodies, you will see endless aspirational rhetoric about ending poverty and starvation with images of smiling African children. These institutions claim they want to end poverty and starvation and they have had decades, big budgets and the influence with national governments with which to do it. Why have they failed?

If you or I had been in charge of the UN since 1970, with access to the politicians and budgets they have, I suspect only a tiny proportion of the world's population would be starving today. Not even with an increased budget, but the same budget. Let's be honest, we could probably achieve it at a lower cost.

The UN has 17 Sustainable Development Goals and, as we saw earlier, has partnered with the WEF which has also adopted these goals. 'Sustainable Development' is a mantra these organisations appear to be quite obsessed about. Indeed, these 17 Sustainable Development Goals are the cornerstone of Agenda 2030 which all 193 United Nations members signed up to in 2015. Goal 2 is 'Zero Hunger'.

How can any right-thinking person rely on the UN to deliver its 17 lofty objectives when it has failed so miserably to tackle the most important thing of all for decades? Whatever one believes about world population, even if he rejects all the arguments in this chapter and chooses to believe the world is at risk of overpopulation in the future, it does not alter the fact that the UN has done a terrible job at reducing starvation and famine to date.

Why should we trust the UN? This is a body that has shown so little desire to ameliorate world starvation it leaves one wondering about their willpower. Does the UN want starving people to survive or not?

Try to imagine a coordinated world hunger campaign where millions of children skipped school and attended rallies where cities were brought to a standstill, but all for starving kids in the global south. One which was promoted endlessly by social media companies and PR wizards, with a child leader who was ferried to the UN conference to berate world leaders about their failures, to be played repeatedly on the news.

Sound familiar? The Greta Thunberg campaign is run by global corporations and green washing NGOs funded by billionaires and it is publicised relentlessly. The campaign's message of fear is based on dodgy modelling, not reality. We are constantly told to be terrified of impending catastrophes that never materialise. On the other hand, the hunger issue is historic and real and there are millions of dead bodies to attest to the UN's failure. Spot the difference.

Just as we put a man in space before we put wheels on suitcases, one suspects we will see the world mobilise to deliver vaccines to stick-thin children before they deliver food to them. If the senior executives of the UN collectively think the same way as Sir David Attenborough and the 'sixth mass extinction' obsessed environmentalists, it might explain their lamentable failures. If a person believes that humans are a 'plague on earth', presumably that person wants fewer of them to exist.

Thomas Malthus' warnings about the world population outstripping the supply of food have been completely disproven. His simplistic argument supposed that the growth of the food supply followed an arithmetic path while population growth followed a geometric path. But he underestimated the ability of humanity to adapt and innovate. This was an early case of bad modelling. When Malthus made his claim, world population was around one billion and in 2022 we are closing in on a population of eight billion. Had his prediction been correct, this would have never been possible. Malthus got it all wrong.

In 1798, the industrial revolution had not yet occurred. Nor had the discovery of nitrogen fertiliser until the turn of the 20th Century, which facilitated massive food production increases. Even if Malthus made his predictions in good faith – which is not the case as we will see – he had excuses for being wrong. Sir David Attenborough does not. He makes the same debunked claim despite over two centuries of verifiable information showing it to be false.

He even refers to Malthus in his speeches as he did in 2011 when speaking to an audience at the Royal Society of Arts.[12]

"The fundamental truth that Malthus proclaimed remains the truth. There cannot be more people on this earth than can be fed."

Sir David Attenborough was falsely conflating hunger with overpopulation and he overlooked the fact that the world produces enough food for ten billion people and could easily increase that output. Is Sir David Attenborough claiming some people *can't* be fed? They manifestly can.

Amazingly, he can make these claims without being challenged about them on the spot. It is also amazing that one of Britain's so-called national treasures can call humanity a 'plague on Earth' without there being uproar.

2.4

However, Resources Will Run Out... One Day

Proponents of the overpopulation myth might counter with something like 'even if we accept that there is enough land and enough food, we are still going to run out of *resources*'. They are usually referring to oil, coal and gas or 'fossil fuels', for want of a better expression.

First things first. Malthusians do not get a pass on having been wrong for 220 years. They should probably pause to reflect on having been horrendously wrong for so long and consider the possibility that they might still be wrong. Secondly, of course, we will run out of resources one day. You would have to travel a long way to find someone who disagrees with this notion, although there are still some abiotic oil proponents who believe oil self generates in the Earth's crust.

Where Malthusians go wrong is believing that the problems are more urgent than they are. They think the issue is so acute that we should be taking steps to depopulate right now. Granted, endless growth on a finite planet is not possible but this is not an argument to commence culling swathes of humanity, aborting the unborn, sterilising poor people en masse and taking the steps that some depopulation advocates recommend.

If new supplies of fuel are not discovered we will run out, although there are sure to be a few false alarms and new finds along that road. When we finally do run out, what would happen?

People would start to die in larger numbers due to poverty and cold weather. But this would not happen as a hard stop or 'cliff-edge' scenario where one day everyone has access to fuel and then overnight we suddenly do not. Wealthier people would hoard scarce resources and poor people will run out sooner. It would be a long time before the last scrap of coal or litre of petrol is finally used. The die-off will have begun long before the last unit of fossil fuel is consumed. (This is assuming there will be no replacement energy source to save us, which is not necessarily true.)

We are a resourceful species with innate survival instincts. These instincts are hard-wired into our genetics. We are quite a hardy lot traditionally and we have had to be to survive to the present day.

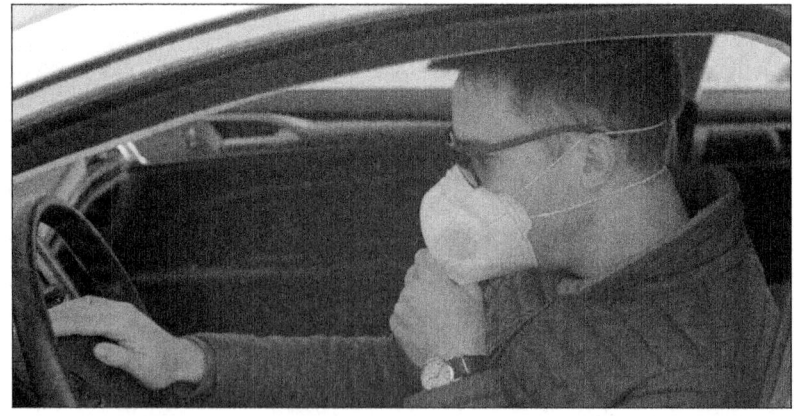

Not you.

Over time, people would be forced to leave the cold areas or perish. Once Scandinavians ran out of wood to burn in the winter, some of them would head towards the equator. That's a big upheaval. Throughout this transition, people would naturally have fewer children because there would be fewer resources and people would be focussing on personal survival. Population would fall naturally.

The point is that there is no need to start that process deliberately.

To the argument that we need to take better care of the planet, undoubtedly we do. But we certainly do not need to pre-emptively dispose of billions of people to achieve that. Most of the destruction of the earth's habitat and pollution is caused by a tiny number of corporations. For instance, when a company like Dupont is dumping dangerous chemicals into rivers and causing illnesses, deformities and deaths via a poisoned water supply, it is not because of overpopulation. Virtually all of us have traces of Teflon in our body, a substance that has no business being there. An overpopulated world did not cause that to happen. Negligent and criminal behaviour of a greedy company did.

Who are the Malthusians to decide we need to depopulate in any case? The person who believes he has the right to tell a people, or a country, or a whole planet of people what the population needs to be, probably has some sort of God complex. Once a person entertains thoughts about the need to depopulate, he is on a slippery slope. All of a sudden, darkly anti-human policies start to look appealing.

One would hope that the notion we should kill two billion people to 'save the planet' or to 'help the rest of the population' is evidently contradictory. It cannot possibly be grounded in compassion or 'the greater good' or whatever twisted logic put forward to justify this barbarity.

Even the people who think they could depopulate 'humanely' with family planning, education and contraceptives are barking up the wrong tree. They are just wrong. It does not need doing because the world is not overpopulated.

Population growth alarmists would argue, "If one admits that the world will run out of energy and minerals one day, then surely one has to agree that the world must be overpopulated?"

The answer to that is no, at least not currently and not by a long way.

The problem with this argument is that half of the world's population is consuming the better part of zero. The lopsidedness of consumption across the world is incredible. As the World Bank reported in October 2020, there were: "689 million people living on less than $1.90 a day. At higher poverty lines, 24.1 percent of the world lived on less than $3.20 a day and 43.6 percent on less than $5.50 a day in 2017."[13]

The 43.6 percent of world population living on less than $5.50 a day amounts to around three and a half billion people who are consuming a tiny fraction of the world's resources. Are we to blame these people because the world running out of jet fuel and lithium?

If these 3.5 billion people could magically be whisked off the face of the earth, it would make practically no difference to our resources dilemma. The world's population would be reduced to around four billion and still have way over 90% of the resources. At the rates that they consume, we could sustain 40 billion of the 'under $5.50 a day' population and they would consume less than the current population does in total.

How strange then, that we are led to believe there are too many of the very people who consume next to none of the resources. Population reduction strategies always seem to be aimed at third-world countries. When the alarmists warn of overpopulation, they always seem to point to pictures of poverty-stricken black and brown people, as if they are the ones the world needs to be rid of. They appear to be saying that we just don't want any more of those poor people – for the 'greater good' of course. But by definition, the poor people are not the ones consuming all the finite resources. How does this make any sense?

How about we flip it around and theoretically remove the lopsided portion by magic – the first world consumers, the people who tend to think the world is overpopulated? It does not take a genius to realise that the depopulation advocates would be 'better' off culling the billion people who are using 90% of the stuff. After the events of 2020, one begins to wonder...

2.5

Did Oil Just Peak?

How about those resources though?

This is a question we need to answer if we are to refute the overpopulation myth. Economic growth, technological advancement and prosperity are all correlated to energy consumption, particularly liquid fossil fuels like oil and diesel. It is that simple. If oil were to run out and there was no replacement source of energy, it would be game over for economic growth and not long after that, population growth. If there were only a few years' worth of oil left, it would be very difficult to claim the world was not overpopulated today.

How much fossil fuel do we have left?

You would be surprised how difficult it is to answer this question. Given all the scare mongering over the past sixty years, one might believe this would be something that is known with a relatively high degree of accuracy. This is not the case. Parties with vested interests have incentives to obscure and flat-out lie about their holdings.

For instance, in 2004, Royal Dutch Shell was fined a record then of £17 million in the UK and $120 million in the US for misstating its oil reserves.[14] Shell overestimated reserves by around 23%, which kept its share price higher than it would have been and failed to correct the exaggeration for three years.

Are we simply to trust the Kingdom of Saudi Arabia when it claims to have 266 billion barrels of oil in reserve? During the 1980s, Saudi Arabia, which had claimed there were 166 billion barrels in reserve at the time, increased that number overnight by 90 billion barrels for reasons that are not entirely clear. Independent audits were not permitted. Thirty years and 10 million barrels per day later, Saudi Arabia still claims it has 266 billion barrels in reserve![15] Whatever new finds Saudi might have enjoyed, surely it does not add up to the last 30 years' 100 billion barrels extracted.

There are other reasons to be suspicious. Why would the Kingdom seek a partial float of its national oil company Saudi Aramco if it has such large reserves? Surely that is just giving value away to outside investors. Why the need to raise cash at all if one is dripping in money and oil? The Kingdom has been shedding its hoard of US treasuries in recent years[16] and is attempting to diversify away from the oil business, which makes one wonder if it possesses as much oil as it claims.

The question of Saudi reserves has intrigued investors for decades and it is easy to see why people are sceptical. Other participants might have reasons to underestimate their holdings rather than overestimate them. One can understand why the government of a small country might prefer to keep news of a big oil discovery under wraps.

As we know, there are powerful forces at work promoting a slew of green agendas. The billionaires and the politicians they sponsor are actively pushing for 'renewable energy' projects which, despite being neither renewable nor environmentally friendly, means moving away from fossil fuels.

When Mark Carney is telling businesses they *will* go bankrupt if they do not submit to his unelected friends' newly created environmental disclosure requirements, it isn't surprising this might hasten the departure away from fossil fuels. The corollary is that the companies he deems to be green-friendly will prosper, through friendlier lending terms and government subsidies.

For these reasons, it is entirely plausible that big oil players wishing to transition to the new regime and grasp some of that green cash on offer might understate their reserves. There likely exist a small number of well-connected and well-informed people who have a very good idea how many decades' worth of oil, gas, and coal we have left but these are not the sort of people who mix in the same circles as us. This ignorance does not seem to bother the Malthusians though.

Their argument goes something like this: fossil fuels are a scarce resource and they will run out one day. They are being depleted by humans. This means there are too many humans so we either need to depopulate or stop having more children.

The fact is they do not even know how long these resources will last which makes their itchy trigger fingers pointing at the head of humanity all the more distasteful. Imagine Mrs Malthus coming home one day in 1798 to learn from her husband, Thomas, that he has just slaughtered their youngest child.

Mrs Malthus: "Why did you do that?"

Mr Malthus: "We were inevitably going to run out of food."

Mrs Malthus: "Well how long until we run out?"

Mr Malthus: "I do not know because I did not look in the cupboards. It is just inevitable."

This is another area where Malthusians do not get a pass. If they insist we desperately need to act because resources are scarce and 'action' entails any step to bring about a reduction in the population, they should have the intellectual honesty – not to mention human decency – to learn exactly how scarce those resources are.

Malthusians have been sounding the alarm with oil depletion in the same way they assured us we would not be able to produce enough food. They have been warning us about oil running out since at least the 1950s when Marion King Hubbert developed his 'Hubbert's Peak' theory, which essentially became the theory of peak oil.[17]

Hubbert predicted that US oil production would peak in the 1970s, that we would reach peak oil around the year 2000 and decline from then on.

Hubbert was a geologist working for Shell Oil who seem to have got it wrong again. To be fair to Hubbert he was not the one screaming, "We're going to run out of oil; HELP, HELP!" at the top of his lungs. Those sensationalist claims were repeated by others for a variety of their own reasons but the reality is that, more than six decades later, world oil reserves have *increased*. British Petroleum's World Energy Review of 2020, estimates there are 1.73 trillion barrels of what they call 'proved reserves' – amounts that can be extracted 'under existing economic and operating conditions'.[18]

When it comes to peak oil, the boy has cried wolf as many times as the climate change lobby. Could the announcement that Shell made in February 2021 be yet another example? *The New York Times* wrote on 12 February 2021:[19]

"With an announcement yesterday – buried in fine print – that its total oil production had peaked in 2019, Royal Dutch Shell made a bold statement about the end of the oil age."

The author says the details were 'buried in fine print' as if he spotted them by diligently scouring reams of documents, yet CNN, *Fortune*, *The Daily Mail* and other big media outlets all reported the story on the same day. When the media collectively reports the same story at the same time it is usually because there is an agenda. Shell was not making a quiet disclosure. This was intended for mass public consumption.

Perhaps a better explanation than the *New York Times'* diligence is the fact that Shell's CEO, Ben van Beurden, made a public statement announcing 'peak oil'. In his statement, he said that oil had peaked, Shell's production would fall every year by around 1-2% and the business would be carbon neutral by 2050. This hardly required journalists to scour the fine print.

Even though he forecast a 45% drop in output by 2050, Shell will still be producing more than a million barrels of oil a day in 2050.[20] One wonders how it is possible to be 'carbon neutral' while churning out a million barrels of oil a day.

A suspicious observer might look at Shell's track record and view this announcement as yet another 'we've hit peak oil guys!' story that isn't true.

It is understandable that some people flat out reject the theory of peak oil. Many who reject the notion that peak oil is just around the corner believe that oil companies have plenty in reserve but understate these amounts to project scarcity and keep prices high. We have seen how relevant parties have incentives to mislead to both the upside and the downside and oil companies may tell lies from time to time. However, we do not even need to deny peak oil to assert that the world is not overpopulated today.

Let us do a quick 'stock take' to estimate what fossil fuels we have in reserve. Here are the author's best estimates and we should qualify these as 'guesstimates' and double underline that categorisation, for who can we trust?

With at least a trillion tonnes of coal, the world has more than 150 years before coal runs out and probably longer. This estimate is the least controversial and almost certainly a conservative one.

The British Petroleum review of 2020 estimates the world's gas reserves stand at 7,000 trillion cubic feet and, if that is true, we should be OK for another 50 years. Notice that proven reserves have increased by about 50% in the past 20 years despite increased consumption. Explorers are constantly discovering new gas fields and we can expect more to be discovered in the future. An estimate of 50 years gas remaining would be an absolute minimum and realistically we have a lot longer than that.

Then there are methane hydrates – methane gas that is trapped in ice crystals in permafrost or beneath ocean floors, also known as 'fire ice'.[21] Large deposits can be found on the edge of continental shelves where the seabed drops into the ocean floor. It is estimated there could be three trillion tonnes of these hydrates – which would last for 1,000 years.[22]

Hazards abound extracting methane hydrates from underneath sea beds, but if a way can be found to extract them without explosions causing submarine landslides and poisoning the oceans, there could be an abundant supply of methane gas.

By far the most important fossil fuel is oil. Oil and other liquid derivatives of oil, power the transport systems that run the world economy, particularly shipping. Taking British Petroleum's recent estimate of 1.73 trillion barrels and dividing by the 100 million barrels per day the world uses, gives us about 48 years' worth of oil remaining. However, it is safe to say that figure of 100 million barrels per day will be falling, even with a slightly larger world population (peaking at around 9 billion, say).

The world is heading for a calamitous economic slump – not your typical recession but something more like a 1930s slump – and this will reduce the demand for oil for a long time. A raft of green initiatives is being implemented internationally and these too will reduce the oil demand, likely permanently. At 80 million barrels used per day, we would have 60 years of oil remaining, at 60 million barrels per day we would have 79 years remaining, and so on. And of course, discoveries will be made, although the 'big ones' are far and few between these days.

The fact is that dozens of countries have already experienced peak oil production and there will come a time when that is true for the others. Shale oil might have 'saved' the US from peak production for a little while but this will only be a short-term fix. The decline rate on shale oil wells is rapid, requiring the constant expensive drilling for new wells. Much of the US shale industry is supported by terrific amounts of debt – a feature of the zero interest rate regime since the 2008 crash – and, when the oil price is under $50 a barrel, production is not even economic. Shale is but a short-term fix and, within a decade, most of the US shale industry will likely be out of business.

When that day arrives, perhaps the years of green rhetoric might bring to an end the USA's habit of plundering smaller countries for their oil whenever it feels like it. Mind, in this day and age it is easy to see the media cheerlead another oil war while guilt-tripping the public for their CO_2 emissions at the same time. Peak hypocrisy and peak stupid will lag behind peak oil. Still, we live in hope.

One suspects we have a bit longer than we are generally led to believe so a conservative estimate would be 50 years of oil remaining. It would be truly surprising if we were to run out before even then.

However – and this is very important to remember – these are the amounts that can be extracted 'under existing economic and operating conditions'.

If there is such a calamitous economic slump that entire industries are driven out of business then energy demand will crater. No one is going to go to the bottom of the ocean floor to extract oil if they do not have a buyer to sell it to, or if it costs more to get it out than they can hope to sell it for. Remember the example of a notional 100 billion dollar oil field located deep underground. It would cost more than 100 billion to extract, even if it were possible. It all comes down to EROI, energy returned on energy invested. There will always be a 'last barrel' of oil but at some point, it just is not worthwhile extracting it because it would take more energy to reach it than would be gained.

2.6

Abiotic Oil

We ought to mention the theory of abiotic oil. This is something of an abandoned theory but it is worth a brief mention if only for its quirkiness.

The abiotic theory of oil claims that oil is self-generating from deep underground, as far down as the earth's mantle. Were this true, then supplies would be limitless but, being so deep down underground, it is not possible to retrieve with existing technology. The oil would only become available once it works its way nearer to the surface through cracks in the earth.

Abiotic oil is said to be formed from non-biological sources (hence the name 'abiotic'), unlike regular oil which is formed when biomass such as dead plankton, plants and the odd animal change form over millions of years.

(This is why the expression 'fossil fuel' is a complete misnomer. Actual fossils are mineralized remains of ancient plants and animals. Oil, gas and coal just aren't fossils. Perhaps the term stuck because the word fossil is synonymous with something ancient. It's probably too late to try to change it now.)

If oil is produced by abiogenesis, we do not seem to have found any of it yet. We know this because when oil is produced in the usual manner, it is a trivially easy task for scientists to prove it. A mass spectrometry analysis will reveal chemical biomarkers that show it originated from plants, algae, plankton and prehistoric megafauna. For oil samples analysed to date, this is what they have always shown.

Also, the world does not appear to be acting as if there is a limitless supply of oil. Dozens of countries have begun to produce less each year showing that for much of the world, peak oil already happened. China sources oil from all parts of the world to power its economy and has invested heavily in hydroelectric and nuclear as well. If there was an inexhaustible supply of oil, China would not be doing all of this.

The abiotic theory of oil sounds fanciful but there are serious physicists and geologists with far more experience and brainpower than the likes of us who sincerely believe it. In defence of the theory, we know that some hydrocarbon fuels can be created without the need for matter that was once alive. Methane, for instance, can be produced from volcanic reactions and it is also known to exist on other planets and moons, particularly Saturn's moon Titan, where there are huge lakes of the stuff. Although methane is not oil, it is a hydrocarbon and this shows fuel sources do not have to be the product of formerly biological life forms.

Abiotic oil is a tantalising prospect but we will just have to reject this theory until we see better evidence for it. Let's not completely close the door though. If somebody can prove it – and show for certain that the oil didn't just seep from a nearby regular oilfield – we should adjust our opinion in a trice. If this ever happens, it will be a happy occasion because it means we won't be running out of oil any time soon. Malthusians would probably regard this as bad news because there would be one less reason to depopulate.

2.7

Other Sources of Energy

The overpopulation myth is such a lie that one does not even need to rely on vaunted green energy 'solutions' coming good, which they most certainly will not.

Wind and solar are both intermittent and therefore unreliable sources of energy. Fossil fuel replacements are required as a backup in the periods they are not available; for instance, with diesel generators, as is the case in the UK. Also, there are huge sunken fossil fuel costs when launching these projects and they generate large amounts of environmental waste, such as the stacks of turbine blades which simply go to the landfill after 15-20 years.

Wind and solar are neither greener nor cheaper and the vast amounts of money being ploughed into these projects will prove to be a waste, ultimately only benefitting the recipients of government subsidies, paid with public money of course. In their current forms, wind and solar are a bust. Hydroelectric power is a better option. Hydroelectric stations constructed with millions of tonnes of steel and cement require an even more mind-boggling initial fossil fuel outlay, but once established these structures last a very long time.

A rather obvious solution to our energy problem is nuclear power, which only accounts for about 10% of the world's electricity today.[23] Decades have been wasted already due to fears over the safety of nuclear energy. As destructive as incidents like Chernobyl and Fukushima were, the overreaction to these events has delayed many advances in nuclear technology which would have benefitted humanity. Germany decided to phase out her entire nuclear industry after the Fukushima disaster in 2011, forging ahead with so-called renewable energy projects instead. It will be a decision Germany regrets and already some of the highest electricity prices in the world are to be found there due to this shift.

Nothing terrifies the public more than an invisible threat and nuclear radiation certainly fits that bill. For a long time, we have been conditioned to fear nuclear power as if it were as dangerous as a nuclear bomb itself. Witness the recent Netflix documentary series 'Chernobyl' for an expensively produced reminder of the dangers of nuclear power. It's almost as if there was an agenda to have us embrace green energy.

The good news is that nuclear energy capacity is on the rise. World leaders, Russia, are at the forefront of this progress and are busy exporting their new fast reactor technology. Dozens of new projects are planned or already under construction in China, India, Belarus, Egypt, Finland, Turkey, Hungary, Bangladesh and South Korea.[24]

Even more encouraging are the rumours that Russia's state-owned atomic energy agency, Rosatom, is close to finding a method of converting Uranium 238 to Plutonium 239 at incredibly efficient ratios never achieved before.[25] If true, this could mean thousands of years' worth of energy. Presumably, we will find out at some point if the boffins at Rosatom managed to crack this problem. (Whether Russia is willing to share this technology with the US and Europe after being pathologically demonised for decades is another question.)

This is not intended to be a foolproof or definitive stock take of the world's future energy sources. We are merely countering the overpopulation myth. Whether the Russian scientists crack this problem or not, they will not be 'saving the world' from overpopulation.

This is where the Malthusians make their single biggest error: they perpetually underestimate human ingenuity, which is ironic given they consider themselves to be the enlightened ones.

The possible Russian breakthrough is just one of limitless possibilities. Who is to say man won't invent a new energy source completely? Some kind of zero-point energy, a cold fusion that works, synthetic oil perhaps?

It takes millions of years for diamonds to form in the earth's crust but we have known since the 1950s they can be manufactured out of carbon in the laboratory in a few days. A million pounds of pressure per square inch while heating to a thousand-odd degrees and Bob's your uncle! A shiny diamond can be made indistinguishable from the real thing. Could a similar process work for the carbon-based fossil fuels that also took millions of years to form under pressure? Granted, expending that amount of energy would be counterproductive but now consider that some bright spark has come along and worked out how to make diamonds in the lab *without* the heat.[26]

Synthetic oil might sound ridiculous – and it probably is – but the point is human ingenuity is limitless. To do as the Malthusians do, ie plot a graph of arithmetically growing food production against geometrically increasing human population, point to the gap and declare 'too many humans' is a grotesque oversimplification and an insult to humanity.

2.8

Taking Stock

If we accept that Shell's peak oil announcement in February 2021 is roughly true for the whole world, that shale oil is not going to 'save' us for very much longer at all, and that the abiotic theory of oil is still just a theory, we still have at least 50 years' worth of oil, 50 years' worth of gas and 150 years' worth of coal remaining.

Due to conflicting information, vested interests and incentives to mislead, we cannot know for sure but these are extremely conservative estimates and we almost certainly have longer than stated. Even with these reserves and a population of 7.8 billion heading to around 9 billion, the world is still not overpopulated. This is true even when you throw in the well-justified assumption that renewable energy projects like wind and solar will prove to be failures. There are three good reasons we can be confident in saying that.

First, we have potentially hundreds if not a thousand years of methane hydrates available to us and methane is the cleanest fossil fuel of all. We can expand nuclear power production and possibly enjoy quantum technological advancements in both output and safe disposal of nuclear waste. We will probably find more fossil fuels as well, particularly in the Arctic.

Extreme Malthusians would probably say that seeing as we will run out in a thousand years, we might as well commence culling the herd today. We will certainly physically run out of fossil fuels one day and we can rest assured that nature will take care of matters when we do. But this is hardly an argument to bring the process forward.

Second, for everything negative said about current solar power technologies so far, there is very likely a clever solution to be found in solar power. Everything we eat, be it plant or animal, began with energy beamed to Earth from the sun which allowed plants to grow through photosynthesis. The same applies to fossil fuels which, for the most part, were once living organisms as well. When you drive a car to work to go to an air-conditioned office and use a computer, every joule of energy was ultimately provided by the sun. Now, consider that the Earth receives less than one part in a billion of the sun's energy.[27] All that energy being emitted, free, untapped. There has got to be a solution.

It would be the greatest and probably most expensive engineering feat in human history but even harnessing 10 parts per billion of the sun's energy would service the world's energy needs for millennia. The point is that human ingenuity has lifted us out of the Stone Age and into the internet age. It is one long story of technological advancement and let's not be forgetting that necessity is the mother of invention, something that is lost on the Malthusians.

Third, as we are about to see, the entire developed world is facing a different problem to overpopulation. Falling population, not overpopulation, is going to be the problem faced by most of the developed countries in the world in the coming decades.

2.9

The Under Population Problem

Population growth is a simple enough thing to measure over two points in time. The difference in headcount between two dates is equal to the number of people born minus the number of people who died during that period (ignoring net immigration). This simple measure does not give us much of an insight as to what will happen in the future though. For that, we need to look at the TFR – the total fertility rate.

The total fertility rate does a much better job of predicting future population levels, which is critical if we are to make sweeping statements such as 'we are at great risk of overpopulation'.

The total fertility rate refers to the average number of children each woman has during her childbearing years. It is not an observable number but an estimate based on the combined rates that women have children at for each age group between 15-49.[28]

A rate of 2.1 is the number required to keep the population constant.

Intuitively this number feels as though it ought to be 2.0. Seeing as there are roughly the same number of men as women if every woman gave birth to exactly two children the population growth would net off to zero when the parents died.

In reality, some women are unable to have children so we need a little over 2.0 from the others to make up the difference. Also, there is a mismatch between the number of men and women in some countries – especially India and China where infanticide has disproportionately afflicted the female population. Currently, India has 1.08 males per female, which amounts to around 50 million more men than women. (In Kuwait there are 1.38 males for each female.[29] The mind boggles.)

Anyway, 2.1 children per woman is the average needed to keep the population steady. If the number falls below 2.1, then at some point the population in that region will begin to fall. Just how long it takes to fall depends on the existing demographics. The median age in Japan is 48 and in Niger, it is just 15. Clearly, it would take longer in Niger for its younger population to die off than it would in Japan, despite the lower life expectancy in Niger.

So what do countries' total fertility ratios tell us?

Anybody who has looked at these statistics will be struck by the lack of babies being born in western and developed countries.[30] If you have never examined these figures then get ready to be surprised.

In the United States, the total fertility rate was 1.78 children per woman, the United Kingdom 1.75, France 1.85, Germany 1.59, Canada 1.53, Australia 1.83, Japan 1.37. All way below 2.1.

Southern Europe looks as though it is positively dying with Spain registering 1.33 births per woman, Greece 1.30, Portugal 1.29 and Italy 1.33. In 1995, Italy was the first country in the world to have more people aged over 65 than under 15. Today there are dozens of countries like Italy.

With similarly sized populations, China and India together account for almost 3 billion people so their rates will have an important effect on the whole. India's rate is probably lower than most would expect at 2.24, but China's even more so at 1.69. Combining India and China, we get a rate of 1.96, which again is below replacement. India has a median age of 28 and China's is 38. That will continue to rise at these birth rates and China will slowly become a country of old people, just like Japan.

That's at least half the world accounted for, and so far we are way below replacement levels.

Russia (1.82) and Brazil (1.74) can also expect to experience falls in their populations. Even Bangladesh's fertility rate is lower than you might think at 2.05 (again, below replacement). Mexico's population looks as though it will stabilise at its current level of around 130 million. It has a fertility rate of 2.14. In the 1960s, it was almost seven!

For South Korean kids, the trend is not their friend. Or their brother. Or sister. Their rock bottom rate of 1.11 children born per woman means millions of South Koreans will grow up without siblings. One hopes for their own sake that South Korea does not get into a war with North Korea (1.91) in a few decades: there might not be enough people to fight it.

All this said, the world population is still trending upwards. So which countries *are* having all the kids?

There is a lot of growth in Pakistan's population, with 221 million people already and a fertility rate of 3.55 children per woman. But to find the most spectacular rates of childbirth, look to Africa.

Nigeria is easily Africa's most populous country with 210 million people. Its fertility rate is 5.42 children per woman. Ethiopia has the second largest population at 117 million and a fertility rate of 4.3. Egypt has 104 million people and a rate of 3.33.

Continuing in descending order of the most populous countries we have DR Congo, with a 91 million population and a TFR of 5.96, then Tanzania 61 million (TFR is 4.92), South Africa 60 million (2.41), Kenya 55 million (3.52) and Uganda 47 million (5.01).

(Note that these population figures[31] will go out of date very quickly with the rate they are growing.)

Most African countries have a fertility rate of over 4 – enough to make the doom-mongers weep – but believe it or not, those rates used to be even higher. Niger, for instance, has the highest fertility rate in the world at 6.95 children per woman. But this is the lowest it has been since at least 1955. It was 7.90 in 1985.

So what is the total fertility rate for the whole world?

That depends on which data you use. The UN estimates the figure to be 2.47 but the most recent estimate according to *The Lancet's 'Global Burden of Disease Report'*, which examines birth rates across every single country, is 2.31.[32] As *The Lancet* report was issued in October 2019, (the most recent version) this figure will already be out of date. The true figure will likely be substantially lower due to the calamitous falls in birth rates across the globe in 2020.

Staying with 2.31 for now though, Africa's very high birth rates offset the rest of the world's low rates giving us a total fertility rate for the whole world of 2.31. As this is still higher than the replacement rate of 2.1, a rate of 2.31 would mean world population is set to climb. This is assuming things stay as they are, however. Of course, they never do.

In 1965, the average woman was having five children – and this is an average figure for the whole world. Today it is 2.31. Clearly, there is a trend – and that trend includes Africa. As we saw in Niger, a whopping 6.95 children per woman is still lower than the 7.90 it used to be and when you look at the numbers this is the case in almost every African country. Africa is a long way off 2.1 children per woman but there are very few reasons to believe the fertility rate will do anything but continue falling.

When you do this analysis it becomes very clear what is going to happen in the medium term. Populations will eventually start to fall in more developed countries and continue to grow in developing countries, particularly in Africa. All the hype about there being 11 billion people, or even ten billion people, is completely overblown.

For many countries, the real risk is that of depopulation. Italy and Japan are two good examples of countries in demographic decline. By 2030, it is estimated that there will be more people over 80 in Japan than under 15.[33] Japan's population has already started to shrink and that will be the fate for many western nations in due course.

The UN estimates that the population is headed back up to 9.7 billion by 2050, but look at Figure 1 to see the assumption they made to get to that figure.

World Population Forecast (2020-2050)

Year (July 1)	Population	Yearly % Change	Yearly Change	Median Age	Fertility Rate
2020	7,794,798,739	1.10 %	83,000,320	31	2.47
2025	8,184,437,460	0.98 %	77,927,744	32	2.54
2030	8,548,487,400	0.87 %	72,809,988	33	2.62
2035	8,887,524,213	0.78 %	67,807,363	34	2.70
2040	9,198,847,240	0.69 %	62,264,605	35	2.77
2045	9,481,803,274	0.61 %	56,591,207	35	2.85
2050	9,735,033,990	0.53 %	50,646,143	36	2.95

fertility rate predicted to RISE again

Source: www.worldometers.info
Based on United Nations data, Department of Economic and Social Affairs, Population Division. World Population Prospects: The 2019 Revision. (Medium-fertility variant).

Figure 1 - The UN predicts fertility rate will rise again up to 2050

The UN is actually predicting a steady *rise* in the fertility rate over the next 30 years. How can this be?

The document upon which these estimates were made is lengthy, overly complicated and requires advanced knowledge of statistics to read, but frankly, the output is so ludicrous it is hardly worth your time.[34] Who in their right mind believes that fertility, having fallen from 5.02 children per woman to 2.31 between 1965 and 2019, will now bounce back to 2.95 by the year 2050? And this is the UN's 'medium' fertility scenario.

Just look at the falling rates in Figure 2 and ask yourself what the UN can be seeing that persuades them to believe fertility rates are going to rise again?

Year (July 1)	Population	Yearly % Change	Yearly Change	Median Age	Fertility Rate
2020	7,794,798,739	1.05 %	81,330,639	30.9	2.47
2019	7,713,468,100	1.08 %	82,377,060	29.8	2.51
2018	7,631,091,040	1.10 %	83,232,115	29.8	2.51
2017	7,547,858,925	1.12 %	83,836,876	29.8	2.51
2016	7,464,022,049	1.14 %	84,224,910	29.8	2.51
2015	7,379,797,139	1.19 %	84,594,707	30	2.52
2010	6,956,823,603	1.24 %	82,983,315	28	2.58
2005	6,541,907,027	1.26 %	79,682,641	27	2.65
2000	6,143,493,823	1.35 %	79,856,169	26	2.78
1995	5,744,212,979	1.52 %	83,396,384	25	3.01
1990	5,327,231,061	1.81 %	91,261,864	24	3.44
1985	4,870,921,740	1.79 %	82,583,645	23	3.59
1980	4,458,003,514	1.79 %	75,704,582	23	3.86
1975	4,079,480,606	1.97 %	75,808,712	22	4.47
1970	3,700,437,046	2.07 %	72,170,690	22	4.93
1965	3,339,583,597	1.93 %	60,926,770	22	5.02
1960	3,034,949,748	1.82 %	52,385,962	23	4.90
1955	2,773,019,936	1.80 %	47,317,757	23	4.97

World Population (2020 and historical) — fertility rate falling for 65 years

Source: www.worldometers.info
Based on United Nations data, Department of Economic and Social Affairs, Population Division. World Population Prospects: The 2019 Revision. (Medium-fertility variant).

Figure 2 - Fertility rates have dropped for several decades in a row.

Even Neil Ferguson would struggle to keep a straight face explaining this kind of model.

The overwhelming likelihood is that before too long, the entire world's fertility rate will be close to 2.1, ie the world population will be at its replacement level. All it would require is for a few populous countries like Nigeria, Ethiopia and DR Congo to get their rates down to 3.0 and the world rate would fall towards 2.1. There will be great disparities though. Within a few decades, populations will be falling in many parts of the world and rising predominantly in Africa and a few other places.

The world is not at risk of overpopulation and the UN is fuelling this myth with this preposterous model. Let's make a prediction: the world's population will not hit ten billion during the lifetime of any person reading this and if it drops under five billion before 2100, something seriously and criminally wrong will have happened.

The average doom-monger who believes the overpopulation myth and gets their opinions from the television probably has no idea about any of this, or even what TFR stands for. That is one of the most dismaying aspects of the overpopulation fantasy. This information is so readily available.

It is easier to parrot a misanthropic lie about the need to exterminate billions than do a simple internet search. That so many people are willing to do so is disconcerting. How many centuries does Malthus need to have been wrong until these people start adding two and two?

2.10

No, Over Population is Not Causing Climate Change

Sorry, doom-mongers. Climate change is not caused by overpopulation either.

Yes, human beings pollute, degrade, poison, over fish and dump hazardous waste and so on; no arguments whatsoever on that score. But man's contribution to raising the global temperature is practically nil. Eradicating the human population would barely make a difference to the levels of CO_2 in the atmosphere and, as mentioned previously, we have just entered a solar minimum, as NASA announced in September 2020.[35]

Depopulation advocates should probably start worrying about the cold weather in future instead of fear-mongering about global warming and hating on their fellow human beings.

Even if dangerous planetary warming was occurring and even if it was caused by humans – which has never been proven – it would not be because of overpopulation but due to the activities of around a hundred corporations, NATO and the Pentagon. Billions of humans leave next to no environmental footprint whatsoever.

And even if the depopulation and global warming alarmists were correct – which they are not – how delightful of them to advocate a reduction in the human population as the solution to save us from ourselves.

Expensive renewable energies and electric cars are not the answer. The 'solutions' offered by the green lobby will inevitably make matters worse because they will keep people in extreme poverty and inflict poverty on people not already suffering from it. Continued poverty itself can bring about environmental problems, although nothing in comparison to the military dropping depleted uranium from the sky. Subsistence farming can cause deforestation as people clear areas to grow food. This problem can be addressed with increased investment in modern agriculture methods, which would bring higher yields, meaning surplus crops to sell, increased self-sufficiency and less deforestation. Investment in disease control, roads, electricity and clean water would also improve the health of the workforce and induce more investment. Ironically improved infrastructure measures such as these also stabilise populations and reduce family sizes, which is exactly what the alarmists wish for.

Poor Africans already have population control by disease and poverty. They certainly don't need birth control, sterilisation and abortion before fresh water and good sanitation. By keeping the poor in poverty, environmentalists are doing more harm to the environment.

When stricken populations climb out of poverty and leave behind the horrors of high infant mortality, they tend to have smaller families anyway, just like people in the developed nations did.

Perhaps this does not occur to the Malthusians. Maybe they believe people from the third world are not capable of rational thought, as if they were somehow genetically inferior? As we shall see, this is precisely what many of them do believe. These problems are simple to solve, if only there was the will to solve them.

Nelson Mandela once said, "Poverty is not an accident. Like slavery and apartheid it is man-made and can be removed by the actions of human beings."

2.11

At the Risk of Repeating Ourselves

Malthus got it wrong. Completely wrong.

If you were to climb into a time machine, set the dial to 1798, get hold of Malthus and ask him, "Mr Malthus, how many people is too many and when exactly does the brown stuff hit the fan in your estimation?" What do you think he would say?

Between 1650 and 1800, the world's population had increased from around 500 million to one billion. This was the context for his alarmism. How many people would he have estimated roamed the earth in his wildest dreams (or nightmares) by the year 2000? Whatever number Malthus told you was 'too many' and whatever year he predicted breaking point, it is a sure bet we would have comfortably sailed past that year with a far higher population.

Imagine his reaction when you told him that, from doubling in 150 years, the population had risen six-fold in the next 200 years? There were seven billion people in the year 2000.

He probably wouldn't believe you because, according to his calculations, there wouldn't be enough food to sustain that population. Not only did the population climb to seven billion by the year 2000, but prosperity also increased by almost any measure you care to name. (This is not to say there is no poverty because there is.

The point is that Malthus claimed it would not be possible, although perhaps he secretly knew it could and had ulterior, sinister motives).

Malthus' predictions were wrong, wrong, wrong. They couldn't have been more wrong. They have been proven wrong for 220 years and yet the likes of Sir David Attenborough appear on our televisions spouting the same nonsense, embellished with climate change predictions which will prove to be equally wrong over the next 220 years.

It is almost as if there is an agenda.

2.12

Surprised?

It might come as a shock to some people to learn that the world is not overpopulated, particularly those who were certain it was true. Face it, that includes most people.

Just as we don't hear the media reporting about the growth in polar bear numbers or vaccine injuries, it is rare to hear reports of plummeting western fertility rates, low sperm counts and childlessness. These are not narratives the media is interested in.

Once a myth becomes ingrained, people tend to stop questioning it. The shocking part is not the information presented here, which can be easily checked, but the fact that so many people believe the myth *given* the information can be easily checked. Why have so many people never bothered to find out if the world is overpopulated?

Even the more educated types are prone to falling for this myth, in fact, *particularly* the more educated types. This is one of the problems with being 'educated', for want of a better word. The type of person with a university education tends to have opinions on a lot of issues, perhaps because he thinks he is required to. When such a person does not actually research a subject, it is tempting to offer the opinion he believes is common knowledge. With a mainstream education, he is more likely to parrot the surface knowledge he has been exposed to and believe it to be the correct answer.

As we said earlier, the overpopulation myth is a lie propagated by the media, academia, think tanks, talking heads and it is the theme of books and Hollywood films. When people hear the same story from every angle, they will tend to assume it is true. People tend not to bother checking information when all their regular sources are saying the same thing.

Remember the wise man we spoke about at the start of this chapter? The one who said, "It's not what you don't know that's the problem. It's what you know for sure that ain't so."

This quote is often attributed to Mark Twain, but erroneously so, as was done in the film 'The Big Short'. It was a wise man or woman who said it alright – and I don't who that person was – but it is not one of Mark Twain's quips.

Here's one that Mark Twain did say for certain though: "Whenever you find yourself on the side of the majority it is time to pause and reflect."

2.13

Is This Normal?

The vast majority of people believe wholeheartedly in a myth that is patently false and whose veracity is simple to check. This is more than just an intellectual tragedy. It is turning large numbers of people into pathological misanthropes.

Take Erik Pianka for example. In 2006, Dr Erik Pianka, a University of Texas professor, gave a speech to hundreds of students at the Texas Academy of Science where he declared humanity was 'no better than bacteria.' Human overpopulation was ruining the earth, said Pianka and we needed to eliminate 90% of the current population.

War and famine would not be sufficient. Pianka favoured a deadly pandemic and bemoaned the fact that AIDS was too slow a killer and not contagious enough. He veritably sang the praises of the Ebola virus, which leads to a hideous, painful, organ-liquefying death. At this point in his speech, a slide of human skulls was shown to the audience. Rick Pearcey, who witnessed the speech, reported:

"When Pianka finished his remarks, the audience applauded. It wasn't merely a smattering of polite clapping that audiences diplomatically reserve for poor or boring speakers. It was a loud, vigorous and enthusiastic applause."[36]

Pianka was presented with a plaque from the leaders of the Texas Academy of Science in recognition of his being named 2006 Distinguished Texas Scientist.

When Bill Gates gave his infamous TED Talk 'Innovating to Zero' in 2010, he showed a slide of an equation that read:[37]

CO_2 = Population x Services per person x Energy per service x CO_2 per unit energy.

Gates said that if CO_2 was going to fall, then probably, one of these numbers is going to have to get pretty near to zero.

There was an audible sniggering from the audience and a few dirty laughs as people realised the implications of his comment. I'm betting that the people in that crowd were not sniggering at the idea they might have to give up their iPhones and cars.

In any case, Bill Gates' equation is incredibly misleading. It does not represent all the CO_2 but *the amount of CO_2 produced by humans*. As we know, human CO_2 output is only something like 3-5% of the total CO_2 in the atmosphere. Gates and his giggling fanboys could depopulate the entire planet and barely make a dent in atmospheric CO_2.

Pianka also miscalculated. In his presentation, he said, "And the fossil fuels are running out, so I think we may have to cut back to two billion [people]."[38]

Pianka does not know how much fossil fuel is left; he is guessing just like the rest of us. Indeed the US shale oil boom began at just about the same time he was advocating this mass murder and the US even became self-sufficient for a brief period. Yet Dr Pianka can't wait to start liquidising billions of people – literally liquidising them with Ebola. The thing is, even if he was right, even if oil was due to completely run out in twenty years, or ten or even five years, it is still not a rationale for killing people.

We've all heard the refrains: 'humanity is a cancer', 'humanity is the virus', we are a 'plague on the earth' and so on. Brainwashed westerners lap this up, somehow believing that billions of penniless foreigners with no political power are harming their precious environment and need to die. They are too lazy to figure out that this is total nonsense but, even if it were true, it is not a reason to speak of killing billions of people.

I have personally heard people say after a deadly earthquake, "Well there are too many of us in the world anyway." I recall a conversation about the death toll in Syria reaching half a million. "Well, at least it's contributing to the overpopulation problem," came the reply. If a bout of freak weather causes the Three Gorges Dam to burst in China and millions of people die, you can fully expect people to say, "Well there are plenty more Chinese."

These comments tend to come from decent and otherwise kind people. Perhaps they are trying to find a silver lining, trying to make light of the horror with a 'joke'? Whatever the excuse, it is a disgusting thing to hear, made all the worse by the fact they have swallowed the overpopulation myth whole. Ignorance is one thing, but this level of malice is something else entirely.

Most people are OK with the concept of population reduction. Normal people. They probably believe it is just an academic issue but if one internalises this thinking for a year or five years or 20 years what are the consequences?

Probably only 0.0001% of people would be physically prepared to cull 90% of the population even if they were in a position to do so. Yet to speak openly of removing billions of humans from the face of the earth is practically mainstream.

There is a danger that talk like this can boomerang back on us with profound psychological effects.

How can somebody simultaneously hold himself out to be a good person while genuinely believing a billion or two people need to die? Imagine carrying those dark thoughts around with you for years, believing that high death tolls from natural disasters are not such a bad thing.

2.14

A Slippery Slope

When one understands that the world is not vastly overpopulated and can laugh at the hysterical fanatics who claim 'we only have 12 years to save the earth', it is easy to realise that the last thing the world needs to do is somehow lose billions of people.

What about the people who believe the bullshit though? How must it affect their day-to-day thinking?

Presumably, if they believe the planet only has room for five billion people or fewer, then every third or fourth person they meet would need to die immediately to get to their desired population. It is only natural that believers would start to have thoughts about whose heads should go on the chopping block, if only in theory. How about noisy people or people who drive too slowly? (We are assuming killers and rapists are already accounted for if you thought this was an odd place to start. With billions to pick from we need to start branching out.) Trump voters? Ah yes, sure to be a popular choice for some. How about meat-eaters? Some of these vegans are pretty hardcore you know. What about those dastardly 'deniers' who refuse to believe we only have 12 years to save the earth? Am I on their list? Are you?

All spoken in jest of course but, after a while, these thoughts must begin to cross the minds of these people.

Here's a little game to play with your Malthusian friends. The next time somebody suggests out loud that this overpopulated world needs to somehow lose a couple of billion people, say this to them with a straight face:

"It's high time we ritually sacrificed a thousand children. Perhaps they could do it at football stadiums around the country on a Saturday afternoon before kick-off."

They would be rightly horrified, yet we are talking about a thousand deaths and they are proposing something hundreds of times worse than the crimes of Hitler. Of course, most people are not advocating actual mass murder, but even speaking about depopulation with such cold detachment leads us down an insidious path. If the 'excess humans' list is billions of people long, that is an awful lot of people to include.

Another unpleasant side effect for these Malthusian reapers is that they might start to feel less sympathetic to people dying in general. They might even look on the bright side, or what they have been conditioned to believe is the bright side.

If we surrender to the idea that it is too late, that we simply have too many people and too few resources, we can become cynical and end up condoning competition for resources. Go down this path and we are at risk of justifying imperial hierarchy and violence.

The attitude becomes, 'OK, so they lied to us about WMDs, but at least we got the oil. Might is right and long live social Darwinism.' The overpopulation myth cons us into believing that the plight of poor Africans and those who live in the Middle East is inevitable. Subconsciously, we begin to side with the very establishment who are committing these crimes, the same establishment that also wants us dead and has been persuading us to go along with their Malthusian cult for centuries.

We don't even need to look as far as other countries anymore. It is happening at home.

Between April and May 2020, thousands of elderly people died in care homes because politicians made the unbelievable decision to take infected COVID-19 patients out of hospitals and place them in care homes with residents who were not infected. The people enacting these policies had to know what would happen. To compound this folly – if indeed it was 'folly' – they were denied hospital treatment and even family visits to the dying. They shut the doors and left thousands to die miserable, lonely deaths.

There was nowhere near enough outrage at this hideous treatment of our elders. No doubt denial plays a part. People find it hard to accept that other people could be so evil, even in such a clear-cut case. To repeat: the people enacting these policies had to know what would happen.

Perhaps our gradual conditioning not to value human life partly explains this insouciance. What have we become when we stop caring about our own people dying?

When one watches television and sees endless images of littered beaches, dirty landfill sites and heart-rending images of birds smothered in oil, it is easy to get upset. Throw in some shrill environmentalists running around squealing that 'humanity is a cancer' and 'mankind is the virus' and people start adding two and two and coming up with five.

When people start talking humanity down as 'bacteria' they have become misanthropes. Worse, they are misanthropes without a cause. Most of their opinions have been downloaded from the television and the defeatist nonsense repeated wholesale, but they are quite wrong.

Referring to other human beings in these sub-human terms is dehumanising and we all know who was famous for that: the Nazis. Eventually though, when we dehumanise other people we end up dehumanising ourselves. In the past when populations have been compared to animals or lower life forms, an extermination campaign has often followed. Today we have this crazy situation where instead of a particular group being singled out, the whole population is being dehumanised. By itself!

The more people who believe human life is worthless the greater the danger and right now billions of people believe this bullshit.

The world is not overpopulated. Humanity is not a cancer. The fact that this Malthusian myth has not been put to bed and the people perpetuating it are given plentiful air time suggests that certain people want us to believe we are bacteria and should be ashamed for existing (*cough cough* climate change lobby).

Children are taught from a young age that the world is overpopulated. Climate change trauma is inflicted on school kids with almost religious fervour and this misanthropic doctrine has taken hold of the collective soul of two generations now. Youngsters are encouraged to believe that they are a burden on the world and as a result leave school supporting depopulation in the name of saving the planet. This cannot be good for their mental health.

These young people are natural prey for groups like Extinction Rebellion where they are guilt-tripped into a constant state of self-flagellation and encouraged to destroy property and disrupt the lives of ordinary people under the guise of environmentalism. In reality, this nihilistic disturbance only serves the agendas of transnational capital while adolescents are being conditioned to believe it is selfish to bring children into the world and in any case why bother as we only have 12 years to live?

The truth is that they have been lied to. They have bought into a philosophy of death that completely devalues human life. Useful idiots teach this in universities all over the US and Europe. The population control and eugenics agenda is the elitist belief system of the perverse 0.0001% who are so lifeless they cannot conceive of other people's right to exist. Yet it is being propagandised by professors, scientists and government officials who have bought into this madness.

One can almost understand the likes of Prince Phillip wanting a tiny population. It would be impossible for everybody to live like he used to. If everyone picked up a gun and slaughtered 30,000 wild animals we would run out of animals and we cannot all own huge swathes of land.[39]

The people who cheer at Erik Pianka's lectures and snigger at Bill Gates' remarks have no skin in the game. These clapping seals have been convinced to go along with the agenda, oblivious that they are part of the enormous mass of useless eaters targeted by the ruling class. If this was a psychological operation it appears to be working. Behold a generation of radicalized nihilists. And these are the 'educated' ones.

If a person watches someone dropping litter and honestly believes the solution is to wipe out millions of such filthy sinners to save the planet, then frankly he is the one with the problem. Rather a litterbug than a mentally ill mass murderer, and it isn't a stretch to label this condition as a creeping, mass mental illness. There is something monumentally awful about wanting to be rid of a thousand times more people than Pol Pot managed to kill, speaking casually about it as if chatting about the weather, while believing it is somehow virtuous. This can have a corrosive effect on our collective psyche. Maybe that is the whole point.

When a man holds so many other humans in contempt he is in danger of hating all of humanity. The next logical step after that one is that he starts to lose all respect for himself. Eventually, such a person is in danger of becoming a miserable self-loather. Malthusian and environmentalist doctrines are likely damaging the mental health of millions of people and contributing to increasing suicide rates.

And all for a lie. Is it really easier to hate all of humanity than do a bit of research?

2.15

Emancipate Yourself from Mental Slavery

It doesn't help that these views are drummed into people from all angles from the moment they are sentient. Schools, media, universities, think tanks and even religious figures are at it. Small wonder so few question the 'official wisdom'.

If we were taught the truth from an early age, that the world is not overpopulated and that human contribution to the global temperature is almost negligible, the mental health of millions of people might improve ever so slightly. Think about the people you know personally who believe billions need to die and consider that they have been carrying those thoughts around for years on end. Do they seem happy to you? Do they struggle with their mental health or have mood swings? Perhaps they have even done the carbon footprint calculations subconsciously and concluded it is they who 'need' to die to save Mother Earth.

Perhaps a dose of the truth might lift some of the background gloom. It won't solve personal problems or get an individual's own house in order but surely it can't help when it is always metaphorically raining outside the house? This is good news. We don't have to be one of those cheering goons at a Pianka or Gates lecture or subconsciously be choosing people for extermination while going about our daily business.

Here's another quote that is always attributed to Mark Twain but doesn't seem to have been said by him either:[40]

"It is easier to fool people than convince them they have been fooled."

It is common to see people reacting badly to the truth if it contradicts their worldview. People can dig in and double down, even becoming angry at the messenger. Rarely has there been a better reason for admitting one's mistake and saying, "Hey man, I got this one wrong." There's no need to be stubborn.

Something became apparent during the first few months of COVD-19 appearing. Many people exhibited bizarre psychology where they would rather cling to catastrophic projections than examine some very basic evidence and change their minds. One hopes that is not the case here. To cling to this myth requires someone to prefer that billions of people need to die rather than just admitting they were wrong.

When the truth has sunk in, it might be time to reflect on how this mass psychosis came about. At this point, it would only be normal to feel righteous, burning anger toward the psychopaths who have deceived us for centuries because they don't think we have a right to exist.

It is at this point I must confess to a little lie of my own. The world IS overpopulated. It is overpopulated with tens of thousands of psychopathic, humanity-hating technocrats and parasites, not content with their abnormal wealth and with minds full of sick ideas.

But there is room enough for all of them to live on a sustainably developed drone patrolled island where they can have their own Great Reset. They can own nothing and be happy with their rations of Universal Basic Income. They can write papers about diversity and inclusion under the permanent gaze of 24/7 electronic surveillance complete with biosensors, tweeting their enlightened goals to one another to be logged irrevocably on their very own intranet-based blockchain while taking orders from a strangely dressed fellow with a bald head and a harsh German accent, until, one by one, they generously contribute to the easing of the overpopulation problem.

2.16

That Goebbels' Quote Again...

While I'm in the mood for confessions, there is something else. Remember that quote at the start of this section – the famous one from Joseph Goebbels about the big lie?

Yeah, I told a little lie about the big lie. See, this quote wasn't from Goebbels. This is not to say Goebbels didn't say it because he did, or at least words to that effect. When he said it though, he was pointing the finger at Winston Churchill and the lying habits of the English.

This is the exact quote Goebbels wrote in a published article dated 12 January 1941 titled *Aus Churchill's Lügenfabrik* which translates as *Churchill's Lie Factory*.[41]

"One should not, as a rule, reveal one's secrets, since one does not know if and when one may need them again. The essential English leadership secret does not depend on particular intelligence. Rather, it depends on a remarkably stupid thick-headedness. The English follow the principle that when one lies, one should lie big, and stick to it. They keep up their lies, even at the risk of looking ridiculous."

It's not the best feeling to find oneself agreeing with a top Nazi but when it comes to the English Malthusians, Goebbels does have a bit of a point. Malthus, Galton, Attenborough et al have stuck rigidly to the lie for 220 years and frankly, they do look ridiculous to a discerning observer. The above quote has been translated from German so it might not be perfect but just to illustrate the context is accurate have a read of a little bit more:

"There is no point to debating Mr Churchill about English ship losses or the damage caused by German air attacks. He follows the time-honoured British policy of admitting only that which is impossible to deny, then cutting it in half, while at the same time doubling or tripling the enemy's losses. This balances the accounts. The astonishing thing is that Mr Churchill, a genuine John Bull, holds to his lies, and, in fact, repeats them until he himself believes them. That is an old English trick. Mr Churchill does not need to perfect it, as it is one of the familiar tactics of British politics, known to the entire world. They made good use of the trick during the World War, with the difference that world opinion believed it then, which cannot be said today. That is because at the end of the World War British plutocracy believed that Germany would never recover. In part from indifference, but also in part from boastfulness, they made the mistake of telling the world the tricks they had used to defeat the Reich. In the memoirs written by British statesmen, Mr Churchill in particular, one could see that the London plutocrats had no problem lying to high heaven during the war. They were even proud of fooling Germany in so easy and clever a fashion. They revealed their methods. They are not believable any longer. We only need to refer to the World War and note that the same men are determining English news policy as did from 1914 to 1918, and everything becomes clear."

It is clear from this section that Goebbels was not admitting to his own deceits, but Churchill's. Of course, it takes a master propagandist to know one. Goebbels was an accomplished liar in his own right but sticking strictly to the facts, the quote we are led to believe was Goebbels self-incriminating is nothing of the sort.

It is impossible to prove a negative such as this but thinking about it; if Goebbels was such a brilliant propagandist and PR man, would he openly admit to being a liar? He clearly states it is unwise to reveal one's secrets.

Many people have great difficulty admitting they were conned and one expects this is the type of fact they would struggle to digest. Anyone who has a problem accepting this message or who has a problem with the messenger has a problem with the truth. Facts do not care about our feelings.

Those who still want to cling to the version that Goebbels did say it about his own German propaganda need to produce some evidence. It's possible he did say it but the proof of him doing so is mighty hard to find.

We can file this one under 'inconvenient truths'.

2.17

Why Does the Over Population Myth Persist?

"Can't you guess?"

Prince Phillip, Duke of Edinburgh

The answer to this question is not particularly a pleasant one. To put it bluntly, the elites have always considered there to be too many people around. Prince Phillip said exactly as much – and just as bluntly – when asked what the biggest challenges in conservation were during the BBC programme *Prince Phillip at 90*. His reply was unequivocal and gave a great insight into the thinking of his fellow conservationist friends, who are also members of the elite:[42]

"The growing human population. From where we are there's nothing else."

When asked, "Do you have views on what should be done about that?"

He replied rather ominously, "Can't you guess?"

Pushed further he suggested 'voluntary family limitation' as a solution and quickly rejected the interviewer's talk of mass sterilisation. But the point is he and his conservationist friends believe passionately that the world is overpopulated. In 1984, Prince Phillip said the human population was 'reaching plague proportions' when it was 4.8 billion.[43]

One imagines he has always believed this, whatever the size of the population. In 1984, he also used the phrase 'voluntary family limitation' as a means to control population.[44]

Most people are familiar with Prince Phillip's famous quip about returning in the afterlife:

"In the event I am reincarnated, I would like to return as a deadly virus, in order to contribute something to solve overpopulation."[45]

People laugh at this. They probably laugh because they think it was a joke and that the overpopulation myth is true. It is not true as we have seen, but Prince Phillip would indeed have preferred that a few billion people were not present on this planet. This is a man who had four children of his own and claimed he wanted to conserve the environment to protect wild animals, having shot dead tens of thousands of them. On the subject of wild animals he had this to say:

"I don't claim to have any special interest in natural history, but as a boy, I was made aware of the annual fluctuations in the number of game animals and the need to adjust the "cull" to the size of the surplus population."⁴⁶

It is easy to see the appeal of the overpopulation myth to kings and queens, governments and the ruling classes over the ages. It offers a great excuse to explain away inequalities. People who complain about their lot in life can be told that all the poverty and suffering is not the fault of the ruling class but a symptom of too much reproduction. Overpopulation is their get-out-of-jail-free card. This is why they have played the running-out-of-food card ever since Malthus (and no doubt before him) right to the present, with helpful mouthpieces like Sir David Attenborough.

Of course, if the population gets too large it increases the likelihood of an uprising. It is easier to control a smaller population than a large one. In ancient times, leaders would make deals with one another to fight wars until a desired number of the population had been killed. Depopulation advocates are as old as the elites themselves.

More darkly, the ruling classes have always been keen to write off certain sections of the population, or indeed entire populations, as inferior and unnecessary. The overpopulation myth allows such people to hide behind their misanthropy and disgust of the lower classes while pretending to care about the well-being of citizens, or more recently, nature and the environment.

2.18

Intellectual Cover

Such views cannot be broadcast honestly by the ruling class to the public or else there might be an insurrection. This is why the elites use men like Thomas Malthus to provide intellectual cover for their twisted views. Malthus might be portrayed as a scholarly man of the church but he was also employed by the East India Company.

Formed in 1600 under the Royal Charter, the East India Company did a lot of the dirty work for colonial Britain and was the de facto government in India. The company operated a college outside London where handpicked young men were trained to be administrators and work in the overseas civil service. Malthus was hired as a professor to train these men from the inception of the college in 1806 and served until he died in 1834. Thousands of Malthusian trained young men would graduate from the college and cast their eye over foreign lands, believing them to be full of surplus, inferior humans.

Malthus was an elitist who advocated killing off the poor. In his essay *On the Principle of Population* he wrote:

"Instead of recommending cleanliness to the poor, we should encourage contrary habits. In our towns, we should make the streets narrower, crowd more people into the houses, and court the return of the plague. In the country, we should build our villages near stagnant pools, and particularly encourage settlements in all marshy and unwholesome situations."[47]

He also advocated that doctors should not cure diseases:

"But above all, we should reprobate specific remedies for ravaging diseases; and those benevolent, but much mistaken men, who have thought they were doing a service to mankind by projecting schemes for the total extirpation of particular disorders."[48]

The verb 'reprobate' means to disapprove, condemn, or censure. These are words that could accurately describe how the remedy hydroxychloroquine was treated by the media in the year 2020, not to mention those 'benevolent but much-mistaken men' and women who tried to 'extirpate' one 'particular disorder', namely COVID-19. The media crucified them. As to the idea we should 'crowd more people into the houses' you could switch the word 'houses' for 'care homes', insert elderly COVID-19 patients, lock the doors and it could be 2020 again.

The idea of infanticide was not beyond Malthus either:.

"All the children born, beyond what would be required to keep up the population to this level, must necessarily perish unless room be made for them by the deaths of grown persons... To act consistently, therefore, we should facilitate, instead of foolishly and vainly endeavouring to impede, the operations of nature in producing this mortality."[49]

Nice guy.

The management of the East India Company and their talent spotters must have appreciated such messages. The upper classes of the day were worried about running out of large open spaces due to too many pesky humans existing. Today we have Agenda 21. Plus ça change.

One supposes that the peasantry of Malthus' day neither had a high literacy rate nor access to Malthus' work. It doesn't get much more blatant than calling for the death of people's children and that might have backfired if enough of the masses knew he was saying this. Prince Phillip could be quite blunt in his own right but even Mr Foot-in-Mouth knew that you didn't go around saying the things that Malthus said, instead of using carefully chosen phrases such as 'voluntary family limitation'.

Malthus was a disgusting, evil man. His predictions were completely wrong on overpopulation but we should consider the possibility he *knew* his predictions would be wrong. Malthus might have been saying what he realised the landed gentry wanted him to say. Expressing opinions as he did is probably the reason why Malthus was afforded academic status and rank. It was certainly a good career move and perhaps he knew it. It is in a similar way that Michael Mann was made a member of the National Association of Scientists and showered with awards and praise after his completely discredited and disingenuous *Hockey Stick* graph, which he cannot possibly have believed in. Malthus was made a Fellow of the Royal Society in 1818. Both men were rewarded for telling the elites what they wanted to hear and there is nothing new under the sun.

Behind his scholarly and religious exterior was a hired pen for the East India Company, a group of brutal colonists and racist thieves, as well as the world's biggest drug pushers. It is telling that Malthus' ideas on population are still referred to in academic circles to this day. Most people have heard of Malthus but so very few people know what sort of a man he really was.

Interestingly, this is the man Sir David Attenborough chooses to quote.

2.19

They've got a Name for That

Another famous depopulation advocate who attracted the attention of the elites was Francis Galton (1822-1911). He attempted to formalise his overpopulation beliefs into a science. He believed the world was overpopulated with the wrong *type* of person and that these lesser mortals should ultimately be removed from the gene pool.

Galton was a half-cousin to Charles Darwin and, after reading Darwin's work, Galton began to propound the theory that humanity could create a race of beautiful, intelligent people by selectively breeding and removing entire bloodlines from the gene pool. He wrote:

"The question was then forced upon me. Could not the race of men be similarly improved? Could not the undesirables be got rid of and the desirables multiplied?" [50]

'Breed the best with the best and hope for the best' is the motto of the racehorse breeder, but Galton felt it could apply to humans as well. He had been working on his theories for a long time but coined a name for this school of thought in 1883. He called it 'eugenics', which translates as 'good birth'. The elites have long had ideas about social cleansing and selective breeding has probably taken place for millennia, but now there was a name for it. Francis Galton is recognised as the founder of the eugenics movement. Adolf Hitler is probably the best-known eugenicist of the last century but he certainly was not the first.

The history of eugenics from Francis Galton's days to the end of World War II might come as a shock to some, so let's have a history lesson that was probably missing from your school syllabus. During this look back over time, you might notice things that are acutely relevant to what is happening today, so we will punctuate the history lesson by jumping back to the present from time to time.

It ought not to surprise us that Charles Darwin, born in 1809, had studied the work of Malthus. Malthus' work was extremely popular as Darwin was growing up. While it is unsurprising that Malthus' ideas were popular before the passage of time proved his predictions completely wrong, it boggles the mind that such rubbish is still taught to children today, 220 years later and after being even more thoroughly exposed.

Curbs on immigration and locking away the disabled were the kind of policies Francis Galton advocated. Galton believed that the elite should be separated from the rest of society and that the genetic underclass should refrain from procreating or else be regarded as 'enemies of the State, forfeiting 'all claims to kindness'. We do not know exactly what he had in mind by that phrase.

He advocated that 'stern compulsion' be used to dissuade those affected by 'lunacy, feeble-mindedness, habitual criminality and pauperism' from breeding. If you were poor, argued Galton, it was a genetic problem rather than an economic one. Galton was also a profound racist with a particular dislike of black people. He wrote a letter to *The Times* with the title *Africa for the Chinese* proposing that the African continent was given over lock, stock and barrel to the Chinese people. He said that 'the gain would be immense' if they were to 'outbreed and finally displace' the native Africans. All of this is openly admitted by the Galton Institute, an institution that exists to this day and is named in his memory. As the Galton Institute website informs us:[51]

"In 1904, Galton established a research fellowship in eugenics at University College London (UCL), and the Eugenics Record Office was installed in Gower Street. The following year the ominously named German Society for Race Hygiene was created in Berlin by followers of Galton."

Galton's views were catching on, particularly in the wrong circles. When people think of eugenics they tend to think of Adolf Hitler and his blonde-haired, blue-eyed Aryan dream. The movement started in London however, with Hitler just the latest believer in this racist quack pseudo-science.

Surprisingly, the Galton Institute only changed its name in 1989, decades after Hitler had repulsed the world with the concept of eugenics. Previously, the Galton Institute was called the Eugenics Education Society, formed in 1907. (Given current trends for hauling down statues and renaming streets, one wonders how much longer the name 'Galton' will appear in the title of this institution.)

Members of the Eugenics Education Society lobbied Parliament, lectured and created propaganda. The Eugenics Education Society was something of an early-day think tank. It was renamed the British Eugenics Society in 1924 to distinguish it from the other eugenics societies, for instance, the American, Australian and New Zealand branches. The movement was particularly popular in the US.

Galton had many other achievements in his career. He was a pioneer in statistics, made contributions to meteorology, helped develop fingerprinting and founded the field of biometrics, a statistical approach to heredity. Remember that last one.

Francis Galton was awarded a knighthood in 1909. In 1910, the Royal Society awarded Sir Francis Galton the Copley Medal, its highest award. He had been a Fellow of the Royal Society since 1860. It was not the first time a rabid depopulation advocate was feted by the elites, rewarded with grant money, titles and prizes. It will not be the last.

On 11 November 2020, the same Royal Society issued a statement saying it ought to be made a criminal offence to spread misinformation about the COVID-19 vaccine in case baseless fears over safety were to impede the vaccine uptake.[52]

How could the Royal Society have known that fears were 'baseless' when it made this statement? By November 2020, not a single study had been conducted which could conclude that any COVID-19 vaccine was sufficiently safe for widespread or even limited use. By definition, it was impossible to know what the long-term effects would be for a vaccine created in such a short timescale. (Speaking of definitions, it is debatable that the word 'vaccine' is even the appropriate term for these experimental gene-based technologies. Having acknowledged this, we will use the words 'vaccine', 'jab' and 'shot' interchangeably).

The Royal Society also made the completely unscientific statement that an 80% take-up of a COVID-19 vaccine could be necessary to protect the community.[53]

Protecting the community with an 80% take-up would be an incredible achievement given the vaccine manufacturers themselves admit their vaccines do not confer immunity at even the individual level or prevent the spread of the disease. (To be fair, the Royal Society did use the word 'could'. I *could* marry Miss World.)

The Royal Society seems desperate to get vaccines rolled out on an enormous and unnecessary scale and is happy to spread very dubious information of its own while calling for criminal penalties for spreading 'misinformation'. Could it be that this is about more than just a vaccine? On 14 February 2021, the Royal Society published a paper titled *Twelve criteria for the development and use of COVID-19 vaccine passports*.[54]

One hears the words of Prince Phillip and his palpable disgust at the plague of humanity. One sees Prince Charles promoting the Great Reset with all his enthusiasm and the Queen herself is lined up on a Zoom call exhorting us to be vaccinated. Meanwhile, the Royal Society enthusiastically promotes vaccine passports. It is not difficult to read the tea leaves in this progression: vaccines, vaccine passports, biometric identities. Copious technology for biometric identity systems already exists and so does the will to implement it, particularly in elite circles such as the Royal Society. Biometric identities are only a stone's throw away and Galton's dreams are coming one step closer to reality.

If you are ever issued with a biometric identity it is worth remembering who the father of biometrics was – the same man who was the father of the eugenics movement. The motives and desires that underpinned his scientific work were to rid the earth of undesirables. If I am ever issued with a biometric identity I will consider myself to have been marked out by the ghost of Francis Galton.

2.20

Minimum Wages

Do you know what the first proponents of the minimum wage had in mind at the turn of the 20th Century?[55]

Far from being a mechanism to help low-wage earners, it was intended to make them even poorer. Economists of the day understood very well that minimum wage laws would increase unemployment because employers would just hire fewer workers. To make matters worse, there was little, if anything, in the way of social security in those days.

As Thomas Leonard wrote in a 2005 paper: *Eugenics and Economics in the Progressive Era*:[56]

"Progressive economists, like their neoclassical critics, believed that binding minimum wages would cause job losses. However, the progressive economists also believed that the job loss induced by minimum wages was a social benefit, as it performed the eugenic service ridding the labor force of the 'unemployable.'"

Progressive economists believed that job losses were a 'social benefit'. Progressive economists such as whom?

How about Sidney Webb, founder of the London School of Economics and one of the founding members of the Fabian Society, a progressive socialist movement? In his 1912 *Journal of Political Economy* Webb wrote:[57]

"Of all ways of dealing with these unfortunate parasites, the most ruinous to the community is to allow them to unrestrainedly compete as wage earners."

With socialists like Webb, who needs capitalists!

President Woodrow Wilson's Commissioner of Labor opposed subsidies to the low-paid, preferring a minimum wage instead. He knew full well it would lead to their culling from the workforce. In 1910, Royal Meeker argued in *Political Science Quarterly*:[58]

"It is much better to enact a minimum-wage law even if it deprives these unfortunates of work. Better that the state should support the inefficient wholly and prevent the multiplication of the breed than subsidize incompetence and unthrift, enabling them to bring forth more of their kind."

This is the sinister part. As well as causing unemployment, minimum wage laws could be used for other nefarious purposes. When Meeker advocated the state support them 'wholly' it was not because the state cared for them, but to prevent them from breeding further and ultimately extinguishing their family trees.

As Henry Seager, the president of the American Association for Labor Legislation wrote:[59]

"The operation of the minimum wage requirement would merely extend the definition of defectives to embrace all individuals, who even after having received special training, remain incapable of adequate self-support... If we are to maintain a race that is to be made up of capable, efficient and independent individuals and family groups, we must courageously cut off lines of heredity that have been proved to be undesirable by isolation or sterilisation."

It beggars belief to read words 'courageously cut off lines of heredity' knowing they come from minimum wage advocates who considered themselves to be 'progressives'. It shows what the real goals of the minimum wage were, cunningly disguised as assistance. Henry Seager and his ilk would no doubt defend their views claiming that they advocated for increased wages, that special training was offered yet still these people could not find employment and were thus reluctantly isolated and sterilised.

It is hardly 'courageous' behaviour to mandate minimum wage laws to make poor people poorer so that whole family trees can be removed from the gene pool. These progressives really fancied themselves: 'brave' as well as ingenious and of course superior to the 'defectives' they wanted rid of.

Today the vast majority of minimum wage advocates probably believe in good conscience they are trying to help the poorest in society. They are not bad people for supporting a minimum wage but their naivety makes them dupes to be used by people who actually understand what they are doing.

2.21

Speaking of Which

Some of the loudest advocates for minimum wage laws today are the so-called progressives and biggest useful idiots in politics. Two such examples are the Democrats Bernie Sanders and former bartender/current actress/Congresswoman Alexandria Occasional-Cortex, I mean Ocasio-Cortez.

A quick word on Bernie Sanders. Bernie Sanders is not a renegade or a maverick or any such thing the media would have us believe. Neither is he a progressive. He is a machine politician, a sell-out and a fraud. When Hillary Clinton was caught red-handed stealing the Democrat nomination from him in 2016, he took his lumps like a faithful sheepdog and kept his mouth shut. When he was sidelined and overlooked again in 2020, and the nomination fixed in favour of an incompetent and malfunctioning Joe Biden, he did the same thing, encouraging his supporters to send their donations to the war-mongering, utterly corrupt and useless crash test dummy who had supported all manner of racist policies in the past. So progressive!

Anything that Bernie Sanders says should be immediately questioned and his motives should too. The less said about Cortex the better, although she is so bereft of cognitive function it is likely she believes minimum wage laws are socially useful. Perhaps she even believes the Green New Deal is a feasible economic policy.

Minimum wage laws are attractive to voters which is why politicians promise to implement them. But are we to imagine the real goal of the minimum wage ever changed? When privileged elites with several cars and houses are advocating for $15 an hour, we should think twice. And then think harder. Think Universal Basic Income (UBI).

UBI is the economic policy that complements and will likely supersede the minimum wage. It was probably created with similar aims and it is even more attractive to the masses: instead of increasing wages, how about free money entirely, for everyone? The lure of free money is likely to raise suspicion in even the most gullible and when people like Hank Paulson, Elon Musk, Mark Zuckerberg, Jeff Bezos and Bill Gates are all advocating for it, we should be thinking, "Hmmmm."

In a truly capitalist, neoliberal, dog-eat-dog world – which is the one we live in – all this talk of UBI would not be given the time of day. The fact that it is – and relentlessly so – should make us suspicious. Traditionally, when movements that genuinely benefit the less well-off members of society gain traction, they are usually crushed mercilessly. Look at how the trade unions were demonised and brutalised in the latter half of the 20th Century. See how the Occupy Wall Street movement was infiltrated and its members physically beaten down in 2011. That is what we would expect to happen in a world where income is grotesquely skewed and bright sparks pop up with ideas to benefit the poor masses. This hasn't happened with UBI, which ought to be a red flag.

For more than a decade, trial balloons have been sent up in the mainstream media and think tanks to promote the idea of a UBI. Economists who support the policy have been given generous air time to explain how it might work. Why?

Are we to believe the wealthiest most privileged billionaires are advocating a UBI for our benefit? Or because they care about us?

Traditionally, the billionaire class has promoted eugenics and you will be hard-pressed to find a billionaire, a single Silicon Valley CEO or member of any institution with the word 'Royal' in it, who will admit that no, the world is not overpopulated. Traditionally these people would prefer that we did not exist on this earth or, at the very least, stopped having kids. Suddenly, en masse, they support the idea of free money for the poor.

If free government money came with no strings attached then that would be one thing. Who honestly believes that it will?

How does 'no vaccine no UBI' sound? How do you like the idea of a completely surveilled economy where you earn credits for 'good' purchases like a subscription to *The New York Times* but are penalised for buying 'bad' things like beef (because cow farts melt glaciers and don't dare question this holy doctrine on pain of a further reduction)? Understand that when government offers a guaranteed income, before long they will practically own you. Look at Chinese social credit scoring and ask yourself honestly if you believe this could never happen in the west.

We are told the Fourth Industrial Revolution will produce amazing robotic technologies that will render entire sectors of the job market redundant. You can take that to the bank. Most of that technology already exists and it is just a question of rolling it out.

Lockdowns implemented under the excuse of a scary virus that killed a whole 0.05% of the world's population will destroy millions of jobs and entire industries. Flatten the curve? How about flatten the economy! What do you think the real purpose of lockdown was? Because when you do the maths, they sure did not save lives, not overall.

Then there are the green-friendly policies which will only increase energy prices and impoverish poor people further.

Health care legislation in the US did not promote good health, probably because the legislation was written by big pharma. Life expectancy fell between 2013 and 2018 in the US, if you can believe that.[60] This is just about the worst indictment of the US healthcare system imaginable. Obscene insurance premiums and sky-high policy excesses just made people poorer and, evidently, sicker.

Do not fall for the propaganda. Promises of free money, Green New Deals and the empty words from the empty skulled Occasional-Cortex are not going to help the common man. These initiatives are vehicles for change and that change is to deliver power to the people governing you. They will make matters a lot worse, for that that is their design. Never trust a 'progressive' bearing gifts.

It should be obvious really. Big budget policies that do not create wealth are most likely to destroy wealth. This is guaranteed to happen in lieu of sane people taking control of the economy. If things carry on as they are then, before too long, people will be begging for their UBI, begging to have nanny state come and save them. Exactly as planned.

Never forget that the rationale for the minimum wage was to remove the poorest people from the job market and get them into the clutches of the state where they could be marked out for sterilisation and even death. Imagine a UBI with a biometric identity to go with it, which might one day be based on DNA and genetics? Think of the power governments will hold over the people by that point.

UBI has been slowly drip-fed into the public psyche over several years to the point it almost feels inevitable. Like the cashless society and biometric identities, it will be presented as 'progress', as something that was just obviously going to happen. This way, agendas that have been planned for decades can be unveiled and imposed on an unsuspecting public to very little resistance. But make no mistake – putting humanity on the dole is the road to serfdom.

2.22

Meet Cloward-Piven

Most people have probably never heard of the Cloward-Piven strategy and it is still referred to as a 'conspiracy theory' even though it is a detailed plan written in black and white, just like Agenda 21.

In 1966, two professors from Columbia University wrote an article in the Nation titled: *The Weight of the Poor: A Strategy To End Poverty*. What they were ultimately agitating for was a UBI.

Richard Cloward and Francis Fox Piven were very popular with Democrat politicians, particularly the Clintons in later years. You might wonder what a pair of grasping establishment politicians like the Clintons found so interesting in a pair of academics who were calling for an end to poverty. Surely their loyalties and interests lay with their banking sponsors and donors?

Cloward and Piven were a pair of subversives. They were well versed in the tactics of Saul Alinsky, an agitating social justice warrior who quite literally devoted his book *Rules for Radicals* to Lucifer. Saul Alinsky had his admirers in high places though. The Clintons, Barack Obama and David Cameron have all since employed Alinsky tactics in their political careers. Such tactics include how to win arguments dishonestly, playing the man and not the ball, and most importantly, winning at all costs.

The Cloward-Piven strategy was to encourage as many people as possible to claim as many welfare benefits as possible with a view to deliberately overwhelming the system. Further, they would exhort radical organisations to make ever more demands for public services at all levels of government. The idea was that the system would collapse under financial pressure, and Federal intervention would be necessary to assume responsibility for the whole welfare system. They made a prediction:

"By the internal disruption of local bureaucratic practices, by the furor over public welfare poverty, and by the collapse of current financing arrangements, powerful forces can be generated for major economic reforms at the national level."[61]

What this amounted to was sabotage against the US economy. But to what end?

Cloward and Piven declared, "If this strategy were implemented, a political crisis would result that could lead to legislation for a *guaranteed annual income* and thus an end to poverty." (Emphasis added.)

Just as the minimum wage was born from sinister motivations, the Cloward-Piven strategy appears suspiciously beneficial to elitists, despite promises to end poverty and claims that payments would be free of government conditions. What they envisaged was a fully-fledged centralised welfare state, meaning more control from Washington. The more people became dependent on the state, the more control and more power the state would have over them.

Now you see why the upper echelons of the Democrat party would buy into such an idea. Democrats like large numbers of poor and unemployed people to exist because they consider their vote is in the bank, almost as if they owned them. Even if we were to believe that the true goal was to reduce poverty the result would be to reduce vast swathes of the population to becoming completely dependent on the state, like groveling infants. The real goal was to put an incredible amount of power in the hands of the few. Winning at all costs indeed, Saul Alinksy style.

There is nothing generous about governments paying huge amounts of welfare benefits either. After all, the money does not come out of the politicians' pockets but from taxation, borrowing and ultimately printing the lot when borrowers stop lending. While it is certainly true that genuinely destitute people would climb to a subsistence level of living, longer-term, it would ensure millions more were lowered to that subsistence level.

To the naive believers who think the politicians cared about poor people, long-term poverty reduction was never going to be achieved. Quite the opposite. Income might be successfully redistributed for a while but, as always, in the process fortunes would be wasted in the grinding gears of government machinery.

Large-scale government redistribution is like a man trying to give himself a blood transfusion by sticking a needle in his left arm, withdrawing his blood into a cup and spilling half of it on the floor before injecting the rest back into his right arm. Wealth is wasted in layers of inefficient government bureaucracy and it perpetuates as these departments grow in self-importance in the mistaken belief they are indispensable.

This is why people like the Clintons would listen to people like Cloward and Piven, just as kings, queens, rulers and elitists listen to people like Malthus, Galton and David Attenborough.

Look at what is happening in the USA today with stimulus payments. It looks an awful lot like the Cloward-Piven strategy in action, doesn't it? It has been a long time coming but when the UBI arrives, it will probably be with the blessing of the whole political spectrum and indeed much of the populace. The millions made jobless by disastrous government policies will have little choice but to accept their meagre state handouts.

Yet again, we see this patient, softly-softly approach where power is centralised inch by inch without people really noticing. Pure Fabian socialist tactics. This is the long road to collectivism. Never trust a 'progressive' bearing gifts. Never trust a Fabian socialist full stop. The movement was founded by eugenicists.

The original coat of arms of the Fabian Society – a wolf in sheep's clothing.

2.23

Eugenics was Mainstream in the 1900s

It is hard to imagine in this day and age just how popular the idea of eugenics was a century ago. Today it is not the done thing to claim openly that entire groups of people are defective, degenerate or sub-human, but in those days it was fair game. Scientists, politicians and intellectuals of the day lined up to express support for the movement. D H Lawrence, T S Eliot, Julian Huxley, Marie Stopes and Virginia Woolf are just a few examples of high-profile eugenics supporters.

The renowned playwright, George Bernard Shaw, was a big proponent of eugenics and had some extreme opinions on what to do with those he deemed inferior; he wanted to kill them and was not shy to admit it. He was also a supporter of Hitler and Mussolini and a staunch defender of Stalin, so perhaps this should not surprise us.

George Bernard Shaw was a Fabian socialist and wrote propaganda pamphlets for the Fabian Society. He suggested gas chambers as a method to kill off the unwanted. In a lecture to the Eugenics Education Society on 3 March 1910, he said, "We should find ourselves committed to killing a great many people whom we now leave living and to leave living a great many people whom we at present kill. We should have to get rid of all ideas about capital punishment... A part of eugenic politics would finally land us in an extensive use of the lethal chamber. A great many people would have to be put out of existence simply because it wastes other people's time to look after them."

This was reported in *The Daily Express* on 4 March 1910.[62]

He also recommended a death panel should decide who lived or died. In Paramount Pictures footage dated 5 March 1931, which can be found online, Shaw says:[63]

"I don't want to punish anybody, but there are an extraordinary number of people who I might want to kill... I think it would be a good thing to make everybody come before a properly appointed board just as he might come before the income tax commissioner and say every 5 years or every 7 years... just put them there and say, 'Sir or Madam will you be kind enough to justify your existence... if you're not producing as much as you consume or perhaps a little bit more then clearly we cannot use the big organization of our society for the purpose of keeping you alive. Because your life does not benefit us and it can't be of very much use to yourself.'"

Presumably a 'properly appointed board' would be staffed with people who looked like, thought like and acted like himself. No doubt they would be government appointments made by Oxford, Cambridge, Harvard or Yale-educated folk.

It is hard to attribute these views to early senility because he repeated them for several decades. He truly did believe what he was saying and what is most alarming is that he also believed he was acting humanely. On 7 February 1934, he wrote in *The Listener*: "I appeal to the chemists to discover a humane gas that will kill instantly and painlessly. In short – a gentlemanly gas deadly by all means, but humane, not cruel."

Not a cruel deadly gas, but a 'gentlemanly' one. Amazingly the progressives of the Fabian Society viewed themselves as the do-gooders who would create a better world.

As well as rubbing shoulders with the wealthy progressives of the day, George Bernard Shaw was a great friend of Nancy Astor. People who know the history of the two world wars – the *real* history as meticulously revealed by Anthony Sutton, or told by consummate insider Carroll Quigley – will know how influential the Astors were in 20th Century world politics. They truly were members of the elite.

Just as Thomas Malthus was celebrated by wealthy and influential people, so was George Bernard Shaw. He was the first person ever to be awarded both the Nobel Prize and an Oscar. His Nobel Prize was for literature, with the comments 'for his work which is marked by both idealism and humanity.'[64]

To advocate the killing of the poor was still a good career move, a century after Malthus. In 1925 it could even earn a man praise for 'idealism and humanity'.

Winston Churchill also supported the eugenics movement and was the vice-president of the first International Eugenics Conference, held in London in 1912. This was a meeting of minds gathered from across the world where dignitaries met to discuss how best to reduce and ultimately eliminate undesirables from the population. Imagine such a conference taking place today.

The conference was devoted to Sir Francis Galton who had died the previous year. Major Leonard Darwin, Charles Darwin's son, was presiding and Arthur Balfour, issuer of the Balfour Declaration, was in attendance. US interests were represented by a group called the American Breeder's Association and they informed the conference of recent US compulsory sterilisation laws designed to eliminate 'defective germ-plasm'.[65]

2.24

Yes, Imagine Such a Conference Taking Place Today

Perhaps it was just a coincidence, but exactly 100 years later in July 2012, in the very same city of London, the British government co-hosted a conference with the Bill & Melinda Gates Foundation. The conference had the innocuous-sounding title of *Family Planning Summit 2012*.

The website for the event featured pictures of smiling Africans, earnest-looking politicians (with Oxford-educated old Etonian David Cameron presiding) and promises to promote women's health.[66] After all, what kind of person does not support promoting women's health?

Attendees included the American abortion chain Planned Parenthood, British abortion chain Marie Stopes International, and the United Nations Population Fund (UNFPA). Melinda Gates said the aims were to deliver "more modern family planning tools to more women in the world's poorest countries."[67]

These tools and services would be contraception, abortion and sterilisation and as usual, they would be aimed at poor people, usually poor black people. Eighteen of the 24 nations represented were from Africa, giving us a clue who the event's 'beneficiaries' would be.[68] The list of commitments pledged $4.3 billion of donations by 2020, with a quarter of this amount to be donated by the Bill & Melinda Gates Foundation.[69]

The pharmaceutical giants were well represented, with Johnson & Johnson, Merck and Pfizer attending, all set to profit from the policies that would be announced. The other donors were the developed world's governments that have provided virtually all the world's population control funding to date. These developed nations have spent $100 billion on population control programmes in the global south since 1995.[70]

Now, cast your mind back to those falling fertility rates we looked at earlier – $100 billion of population control money will tend to have that effect.

A WHO summary document explained what a successful outcome of the summit would be: "By 2020, we aim to serve a total of 380 million women with quality family planning to delay, space or limit their births."[71]

In other words population reduction, with a kinder friendlier face, promoted by billionaires, big pharma and Old Etonians. The conference capped two decades and 100 billion worth of programmes of ever more aggressive but innocuously presented methods to stop Africans, Asians and Latinos from having children, programmes which continue to this day.

None dared call it eugenics by 2012 – not after Hitler had ruined the movement's name – but it looks an awful lot like the same thing.

2.25

Speaking of Hitler

Adolf Hitler did not invent eugenics or the twisted idea of creating 'racial purity'. He was just the latest advocate in a long line of eugenicists. Hitler particularly admired the advances in eugenics that were occurring in the US. The feeling was mutual, with one American eugenics proponent, Joseph DeJarnett, observing in *The Richmond Times Dispatch*, "The Germans are beating us at our own game."[72]

Leon Whitney, of the American Eugenics Society, remarked: "While we were pussy-footing around... the Germans were calling a spade a spade."[73]

The historian, Edwin Black, has detailed in his incredible research how the German eugenics movement was assisted by their American counterparts with scientific know-how and financial assistance, not to mention 'moral' support of the most immoral variety. As Black notes: "The intellectual outlines of the eugenics Hitler adopted in 1924 were made in America."[74]

By 1924, decades of so-called progress had already been made in the field of eugenics in America. Hitler had been observing closely. He studied American eugenics laws and cloaked his sick views in the pseudoscience of eugenics, encouraging moderate Germans to follow his 'medical' philosophies. To draw a 2020 pandemic parallel, he persuaded his countrymen to 'follow the science'.

In his 1924 book, *Mein Kampf*, Hitler wrote: "There is today one state, in which at least weak beginnings toward a better conception [of immigration] are noticeable. Of course, it is not our model German Republic, but the United States."

Hitler wrote[75] to leading eugenics author, Yale-educated Madison Grant, praising his race-based eugenics book *The Passing of the Great Race*.[76]

Hitler referred to Grant's book as his 'bible'.

Mere ideas do not create genocides, however, or mass deportations or forced sterilisations. For these crimes to occur, popular support and legal backing is necessary, which in turn requires organisation, political will and of course funding.

Money was never going to be a problem because the financial elites have always supported eugenics and population control. In the US, funding came from foundations established by the wealthy industrialists who had become America's richest families. The Harrimans, the Carnegies and the Rockefellers all contributed generously to eugenics programmes and organisations. Known as the 'robber barons' for their ruthless exploitation and monopolistic practices, they now turned their hand to propagandising eugenics, funding scientists and lobbying governments. Even in the 1900s, the elites' dark objectives were to be concealed under the guise of philanthropy.

They employed scientists from Ivy League universities such as Stanford, Yale, Harvard and Princeton to promote race theory and race science and provide intellectual cover for their racist classist beliefs.

In 1904, the Carnegie Institute established the 'Eugenics Records Office' called Cold Springs Harbor Laboratory on Long Island. From here they agitated for changes to the law and lobbied social service agencies and associations. Millions of index cards were created for ordinary Americans as researchers earmarked families and bloodlines for removal. The Cold Springs Harbor Laboratory still exists to this day.

The Harriman railroad fortune helped fund charities such as the New York Bureau of Industries and Immigration to locate Jewish, Italian and other immigrants in major cities to have them deported or forcibly sterilised.[77]

The Carnegie Institute funded a 1911 *America Breeder's Association Report* which proposed a range of solutions for the 'best practical means for cutting off the defective germ-plasm in the human population'. Sterilisation, segregation and euthanasia were all proposed, as can be seen in this picture of page 5 of the report.[78]

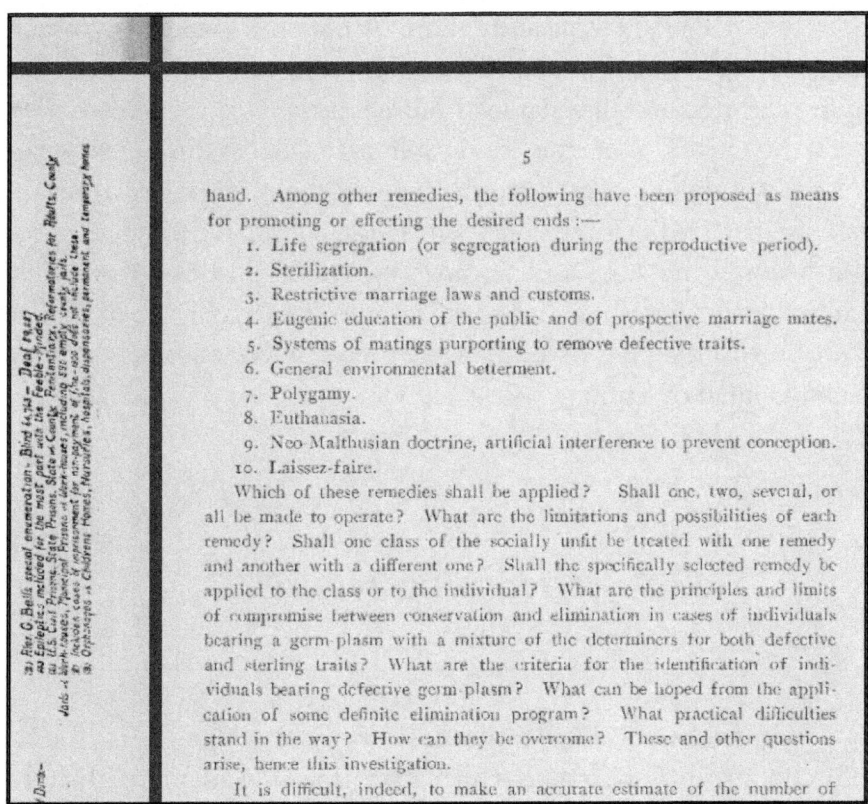

Before long, sterilisation and segregation laws were passed in many US states as well as restrictions on marriages.

In the infamous case of Buck vs Bell in 1927, Supreme Court Justice, Oliver Wendell Holmes, rejected Carrie Buck's plea to prevent her court-ordered salpingectomy to make her sterile. She had been deemed feeble-minded by the state of Virginia. Justice Holmes opined:

"It is better for all the world, if instead of waiting to execute degenerate offspring for crime, or to let them starve for their imbecility, society can prevent those who are manifestly unfit from continuing their kind... Three generations of imbeciles are enough."[79]

Carrie Buck was in fact of normal intelligence and had fallen pregnant because she was raped by a relative of her foster family.

After this decision, thousands more Americans would be forcibly sterilised. As many as 60,000 Americans were prevented from having children and thousands of more marriages were refused.[80]

Twenty years later, the Nazis relied (unsuccessfully) on Justice Holmes's judgement in their defence at the Nuremberg trials. As well as that, Hitler's personal physician Dr Karl Brandt, charged with heading the Nazi euthanasia programme, justified the Reich's policies on the basis they were not unique or original. He introduced Madison Grant's *Passing of the Great Race* into evidence – the book so appreciated by Hitler.[81]

2.26

A Helping Hand

Practical eugenics measures were firmly established in the US before being off-shored to Germany. But it was more than just mutual admiration powering the German eugenics movement. Germany found plenty of support from willing American helpers, with California-based eugenicists busy producing and circulating booklets to German officials and scientists. They even displayed Nazi scientific exhibits in their museums, such as the 1934 display at the LA County Museum for an American Public Health Association meeting. Time and again, assistance came from California. As Edwin Black explains:[82]

"Much of the spiritual guidance and political agitation for the American eugenics movement came from California's quasi-autonomous eugenic societies, such as Pasadena's Human Betterment Foundation and the California branch of the American Eugenics Society, which coordinated much of their activity with the Eugenics Research Society in Long Island. These organizations – which functioned as part of a closely-knit network – published racist eugenic newsletters and pseudoscientific journals, such as *Eugenical News and Eugenics*, and propagandized for the Nazis."

American eugenics did not just lead the Germans by example, but with hard cash. Personal and professional relationships were cultivated with the American's fascist counterparts. The Rockefeller Foundation made a $250,000 award to establish the Kaiser Wilhelm Institute for Psychiatry, whose staff included Hitler's medical henchman Ernst Rüdin, responsible for conducting beastly experiments on Jews and gypsies.[83]

The Institute also undertook extensive research into twins, something that had fascinated American eugenicists who wanted to better understand heredity.

Head of the Kaiser Wilhelm Institute for Anthropology, Human Heredity and Eugenics since 1927, was Otmar Freiherr von Verschuer who set about his work studying twins. He had an infamous assistant called Josef Mengele. When Josef Mengele arrived at Auschwitz in 1943, Verschuer notified the German Research Society:

"My assistant, Dr Josef Mengele (MD, PhD) joined me in this branch of research. He is presently employed as Hauptsturmführer (captain) and camp physician in the Auschwitz concentration camp. Anthropological testing of the most diverse racial groups in this concentration camp is being carried out with permission of the SS Reichsführer (Himmler)."[84]

Mengele would abscond after the war. Theories about his whereabouts and activities after the war vary, but Verschuer's fate is known with certainty. Verschuer somehow managed to escape punishment for war crimes entirely, despite having joined the Nazi party and even though there was solid evidence he collaborated with Mengele by his own admission. He had even written in *Der Erbarzt*, a eugenics journal which he edited, that a "total solution to the Jewish problem" would result from the war.[85]

Yet, despite all this evidence, Verschuer was not tried for war crimes and continued his career in genetics after the war. Perhaps certain people of influence believed his contribution to eugenics merited a free pass. Verschuer was back in contact with the California eugenicists after the war and became a respected scientist, joining the American Society of Human Genetics in 1949 as a foreign correspondent.[86]

'Human Genetics' was the new term for eugenics, which had been declared to be a crime against humanity.

Students of World War II history – the *real* history – will be familiar with Operation Paperclip. Thousands of high-value Nazis were moved to the US after the war and transplanted into important roles on behalf of US intelligence. Many had their records sanitised and were given new identities. In some instances, family members even travelled with them. Of particular interest to US recruiters were Nazi scientists with expertise in rocket technology and chemical or biological warfare knowledge.

Some incredibly senior positions were filled by former Nazis, perhaps the most notable being Wernher von Braun, who headed NASA's engineering programme. Erich Traub, a biological warfare expert, helped to found Lab 257, a secretive military germ laboratory at New York's Plum Island.[87] Traub had been to the US before. He had studied at the Rockefeller Institute in Princeton, honing his knowledge in viruses with training from American scientists. He returned to Nazi Germany just as the war was about to start. Now he was back in the USA, doing what he did best.

Other Nazis made the opposite journey. Nazi General, Reinhard Gehlen, was brought to the US and then sent back to Germany to do the CIA's work. Gehlen was rescued in September 1945 and brought to America with the help of the OSS, the forerunner of the CIA.[88]

Reinhard Gehlen had been in charge of the Russia desk for German intelligence and had a treasure trove of secret files on their Soviet enemies. He would go on to head the Russia desk for the newly formed CIA and when he was sent back to Germany to perform this role he set about hiring his former Nazi colleagues. Reinhard Gehlen had been a consultant on the Final Solution. This was the sort of man who now worked for the CIA.

It was a huge recruitment effort and transfer of dark technology that would help lay the groundwork for the Cold War with the Soviet Union. Even at the time of their greatest triumph, as the Red Army was on its way to Berlin, Cold War plans were being hatched.

Other Nazis would take up senior positions at NATO, including Hitler's former chief of staff, Adolf Heusinger. In 1961, Heusinger would become NATO's chief of staff. Another was General Hans Speidel, who had been Erwin Rommel's chief of staff during the war. After the war, he became the Supreme Commander of NATO's ground forces in Central Europe from 1957-1963. All the while, monsters like Verschuer were left in Germany unpunished. For some of the worst criminals of the 20th Century, it was as though nothing had ever happened.

Wall Street unquestionably funded Hitler's rise to power and major US corporations assisted the German war effort, detailed beyond a shadow of a doubt by the historian Anthony Sutton.[89]

The Californian eugenics crowd had helped the Nazis with their eugenics programme, with a large helping hand from the Rockefellers as well. After the war, some of the worst Nazis would switch sides and go back to work for their American benefactors. Some had even been trained in the USA before the war and were returning students.

Most people today are not aware that card-carrying Nazis were rehabilitated and would go to work in US government programmes and become high-ranking CIA and NATO personnel. This is an inconvenient truth overlooked by most western universities and schools, an oversight that has cost the world dearly. For instance, would both George Bush Presidents have been elected if a properly educated electorate were aware that Prescott Bush was caught red-handed helping the Nazis?

Prescott Bush was known to have collaborated with the Nazis before and during World War II through his work with bankers Brown Brothers Harriman and Union Banking Corporation.[90]

This continued even after the bombing of Pearl Harbour. His German assets were seized in 1942 by President Roosevelt under the

Trading with the Enemy Act. Even though reprimanded, Prescott Bush was not sent to prison nor did it prevent him from becoming a Senator for Connecticut between 1952 and 1963.

Prescott Bush claimed he was not sympathetic to the Nazi cause but he certainly helped them financially and it is inconceivable he was unaware that his business deals were being driven by slave labour.[91] Prescott Bush was also a supporter of the American Birth Control League, the forerunner of Margaret Sanger's Planned Parenthood and was treasurer for that organisation in 1947. Yet another population control advocate.[92]

When one observes the unspeakable horrors inflicted on much of the world by the CIA and NATO in the post-war years and even on American citizens, particularly through medical and mind control CIA programmes, it makes one wonder who could be capable of such evil. An obvious answer to that question would be 'Nazis' – and the people who rehabilitated them.

As author Joseph P Farrell remarked when reviewing the World War II surrender documents: "The German armed forces surrendered, and Germany surrendered. But the Nazi party did not."[93]

Of the many American corporations that assisted the Nazis throughout the war, one in particular warrants a special mention: International Business Machines (IBM). The speed with which Hitler identified Jews so efficiently and accurately so that they could be targeted for segregation, asset forfeiture, imprisonment, forced labour and death has surprised historians. Edwin Black can provide an explanation.

IBM helped Hitler in this complex task, in an alliance that began almost as soon as Hitler came to power in 1933. The computer had not been invented in the 1930s but IBM's punch-card technology had. The President of IBM, Thomas J Watson, established subsidiaries in Poland and Germany to collude with the Nazis and this business continued throughout the war, mainly through IBM's German subsidiary Deutsche Hollerith-Maschinen GmbH, which had monopoly rights in Germany.

Complex solutions were designed to fit the Reich's requirements, including Hollerith Departments at every concentration camp which allowed the Nazis to keep tabs on inmates with IBM's punch-card technology. Edwin Black's research has demonstrated that these activities were ultimately managed from IBM's headquarters in New York. IBM has since made payments to Holocaust survival groups, without admitting liability.[94]

IBM would feature in the story of Bill Gates' rise to riches, a story that is the stuff of legend. Bill Gates got his big break when he licensed his operating system to IBM in 1980. The amazing deal Bill Gates negotiated with IBM features in several documentaries about Bill Gates' life. A deal so amazing it stretches credulity.

We are led to believe that Gates engineered not just a sale, but a deal with lucrative licensing agreements that netted Microsoft billions over the years. We are told that the socially awkward college dropout Bill Gates, with two cents worth of business experience and even less charisma, took on the mighty IBM, a corporate leviathan with access to teams of lawyers. We are supposed to believe that Bill Gates got the upper hand over IBM trading an operating system for $50,000 that he didn't even design himself but bought off the shelf from a competitor.

You are welcome to believe this story but the bullshit smells particularly strong to me. Was IBM incapable of offering the guy $50,000 themselves for his operating system and cutting Bill Gates out altogether? When books are written about billionaires and films are made praising their genius, it is hardly likely to be the unvarnished truth.

Just remember that the meeting with IBM was set up by Bill Gates' mother. Mary Gates was acquainted with the IBM chairman John Opel through other Board memberships and she pulled some strings for the young Bill Gates.

It has been speculated that IBM knew it would fall foul of antitrust laws if it ran its own operating system and so 'outsourced' this role to Microsoft. If that is the case, then Microsoft would be a mere spin-off of IBM. This is pure speculation however and evidence to support it proved elusive to this author.

One thing is for certain though; Bill Gates had a huge helping hand in the transaction courtesy of his mother's connections. The chances of Bill Gates being where he is today in a truly meritocratic, free market and without his parents' help, are touching zero.

And where is Bill Gates today exactly? Having relinquished his Microsoft board membership in 2020, he has emerged to be the world's foremost pusher of vaccines. With no medical qualifications whatsoever, Bill Gates set up in philanthropy, giving away billions while miraculously doubling his wealth and investing copiously in the same companies that governments would buy billions of vaccines from. Gates has been working very closely with the Rockefeller Foundation for years and the parallels between the 'philanthropic' work Rockefeller carried out in education and health and which Gates does today are uncanny. Gates is the largest private funder of the WHO and Dr Anthony Fauci is on the leadership council for the Bill & Melinda Gates Foundation. The potential conflicts of interest are endless.

2.27

Back to the Present – Moderna and IBM

Since those dark days assisting the Nazis, IBM has assisted the US government in the space race (in collaboration with NASA) and provided the US military with a helping hand in several areas over the decades.[95] IBM maintains an unlikely connection to the Pentagon through its dealings with a vaccine company. In 2021, IBM entered into an agreement with Moderna, the manufacturer of a COVID-19 vaccine which uses messenger RNA technology.[96]

On 4 March 2021, IBM and Moderna announced that they would use technologies such as artificial intelligence and blockchain to 'improve information sharing' and 'to address potential supply chain disruptions'. IBM would also be offering blockchain technology to 'help individuals maintain control of their personal health information' and help 'organisations better verify the health credentials of employees, customers and travellers.'

Given the company is in pole position in the COVID-19 vaccine race, certain facts should be known about Moderna.

Moderna's website tells us that the Bill & Melinda Gates Foundation is a 'strategic collaborator', with the Gates Foundation committing up to $20 million back in 2016.[97] Moderna also has very direct links to DARPA, the Defense Advanced Research Projects Agency. Moderna's mRNA vaccines were developed with the help of a $25 million grant from DARPA way back in 2013.[98]

DARPA is the Pentagon's secretive research unit and all sorts of amazing new technologies emerge from here. DARPA projects include militarised robotics and nanotech weaponry, as well as bioweapons and biotechnology such as microchips that are claimed to create and delete memories in humans. These are truly cutting-edge technologies.[99]

The trouble is that everything studied and created by DARPA – as the name suggests – is done with a military objective in mind. One does not usually associate vaccines with military use.

In 2017, concerns were raised when it was discovered DARPA was investing $100 million in genetic extinction technologies. *The Guardian* reported that so-called 'gene driving' can alter the genetic constitution of target populations which can in theory lead to their eradication.[100]

The Guardian's assurances that 'known research is focused entirely on pest control and eradication' are probably intended to alleviate concern. But what about DARPA's unknown research? As to 'genetic extinction technology', think about those words in terms of eugenics.

2.28

Fools Rush In

Money from Bill Gates and the dark side of the US military has funded this vaccine, now being rolled out on a global scale with the help of IBM, even though there is no knowing what the long-term effects of mRNA vaccine technology in humans will be. And all for a virus that is essentially a bad case of flu.

This is the nexus of the military, technology and big pharma meeting at the point of a needle to be injected into the arm of humanity. It was rolled out as part of a project called 'Operation Warp Speed' and as you know, when you travel at warp speed you leave the known universe behind. Regular vaccines can take between four and ten years to be properly tested and approved but the Moderna vaccine was fast-tracked at a speed never seen before, despite being a new technology for human use. Multiple world leaders issued coordinated messages assuring us we all needed to be vaccinated for our own good in April 2020. By December 2020, the Moderna shot was being injected into arms – just eight months later – along with the other hastily prepared COVID-19 jabs. What could possibly go wrong?

It is a struggle to believe that this is all about public health. Recent history has shown that most of the world's elected governments appear not the least bit interested in public health. By granting immunity to the pharmaceutical giants they have signalled whose side they are on and it isn't yours.

2.29

Will Moderna Win the Gold Medal in the Vaccine Race?

Almost as soon as the first doses of vaccines were being rolled out, there was talk of mutant strains of COVID-19, South African variants, Brazilian variants, Kent variants and so on. People were asking whether the vaccines would offer protection against all these new 'scariants' and questions were asked about how long the vaccines would offer protection in any case. Nobody seemed to know, which was not very encouraging but hardly surprising given the speed with which the vaccines were rushed out.

Soon the talk turned to regular 'booster' jabs to keep people updated from all these new scary threats. This is the very thing that the more critically thinking people had been warning about from the outset. We will never be free from COVID-19 vaccines and to keep vaccine passports up to date, one will need regular booster shots. Eventually, the vaccine passport will not just be for travelling abroad either, but to gain domestic 'privileges'.

Anyone labouring under the illusion that COVID-19 will eventually blow over one day should face up to this reality quickly. Unless enough people refuse to comply and the whole scheme implodes, we can look forward to annual shots or even six-monthly shots.

Were one to cast their mind forward a few years to a time where regular vaccines are a normal part of life, there are a couple of good reasons to believe that the Moderna vaccine will become the world's 'chosen' jab.

The Moderna vaccine uses the mRNA technology which Bill Gates enthused about on his blog in April 2020, very soon after

COVID-19 was hyped as this enormous problem. The blog post had the hysterical and demonstrably false sub heading:[101]

"Humankind has never had a more urgent task than creating broad immunity for coronavirus."

This is such a ridiculous claim it does not even merit compiling a list of more urgent world problems. In any case, we have hydroxychloroquine and ivermectin. You can chill out Bill, problem solved.

The technology for mRNA vaccines had been in the pipeline for years and Bill Gates had helped fund its development. Somehow Gates surmised as early as April 2020 that we needed to vaccinate the whole planet for a disease whose lethality was not fully known but early indications suggested was very mild and practically non-existent in children. The medically unqualified but hugely influential Gates determined that this technology was what we would all need. Bill had his hammer and every virus was looking like a nail. Here was a huge clue as to what Bill Gates wanted and he does have a habit of getting his own way. Notice also that there is a pattern where national governments fall into line with the policies that the non-doctor, non-medical expert, computer guy Bill Gates promotes.

Then there were problems with Moderna's competitor vaccines, an utterly predictable scenario given that all previous attempts to produce coronavirus vaccines had failed. The AstraZeneca vaccine was quickly withdrawn or restricted by dozens of governments that were concerned about the injuries it was causing to their citizens, particularly problems with blood clots.

The AstraZeneca vaccine has some 'interesting' ingredients, shall we say. This quote is taken directly from the AstraZeneca product information sheet:[102]

"One dose (0.5 ml) contains: Chimpanzee Adenovirus encoding the SARS-CoV-2 Spike glycoprotein (ChAdOx1-S)*, not less than 2.5×10^8 infectious units (Inf.U.)

*Produced in genetically modified human embryonic kidney (HEK) 293 cells and by recombinant DNA technology.

This product contains genetically modified organisms (GMOs)."

The original HEK293 cell line came from an aborted human fetus. As the website of the same name hek293.com clearly states:

"The source of the cells was a healthy aborted fetus of unknown parenthood."[103]

Lamentable 'Fact Check' websites line up to claim the vaccine contains no aborted human fetal tissue. Strictly speaking, that is correct. The vaccine does not contain aborted human fetal tissue per se. It contains cells from a cell line that *originated* from an aborted human fetus. So that's OK then.

A hint that the Pfizer vaccine might not be the world's COVID-19 vaccine of choice was dropped on the day of the well-publicised announcement of Pfizer's supposed trial success. As the company was making dubious claims of 95% efficacy based on just 170 infections during trials[104] and the media was feverishly pushing this story on the public, Pfizer CEO Albert Bourla was busy selling 62% of his company stock, $5.6m worth.[105] If Albert Bourla believed that the announcement spelled a high watermark for the stock, I am inclined to agree with him.

Lest we forget that Pfizer was issued with a $2.3 billion fine in 2009. This was not the usual civil class action but an actual criminal prosecution. Pfizer lied about the efficacy of its products – some of which caused fatalities – and bribed doctors in a case described by the US Department of Justice as the 'largest healthcare fraud settlement in its history'.[106]

It was just one sorry episode in a very long list of sorry, deadly Pfizer episodes. Pfizer's dark history is littered with penalties and fines, a rap sheet no corporation would be proud of.[107] It is easy to understand why Pfizer would seek complete immunity from redress with its COVID-19 vaccine. It is far from easy to understand why any government would grant it.

The Pfizer vaccine also uses mRNA technology and contains an interesting ingredient: potassium chloride.[108]

This ingredient is used in lethal injections for death penalty convicts, so I will pass on that experimental vaccine too. Potassium chloride helps to induce death by stopping the heart. Although the amount of potassium chloride in the vaccine is orders of magnitude lower than the amount death row prisoners are injected with, the question remains: what is it doing there?

At the time of writing, none of these COVID-19 jabs had been approved by the FDA. They are still in the experimental stage and only have Emergency Use Authorisation (EUA). Do you think the average punter who queued up to be injected is aware of this fact? These people are guinea pigs in a massive human trial. The clinical trial documents state this clearly. Moderna's estimated study completion date is 27 October 2022.[109] Pfizer/Biotech's estimated study completion date is 2 May 2023 [110] and AstraZeneca's is 14 February 2023.[111]

Do you think this information was presented to recipients before they were jabbed? Was their consent for participation in a medical trial obtained? It makes an absolute mockery of the phrase 'informed consent'.

Some big-hearted soul at AstraZeneca picked Valentine's Day as the endpoint for the human trial. By 'big-hearted' you can read 'enlarged', seeing myocarditis is a side effect suffered by far too many of these unsuspecting volunteers. If current progress is maintained then by 14 February 2023 the trial could legitimately be called a massacre.

If you think that sounds hyperbolic, consider that in 1976, the swine flu vaccine rollout was called to a halt after just 53 deaths from 45 million doses.[112]

Now take a look at the US government's Vaccine Adverse Events Reporting System (VAERS) deaths report. The chart below shows all vaccine deaths reported to VAERS over the past 30 years. It isn't exactly subtle.

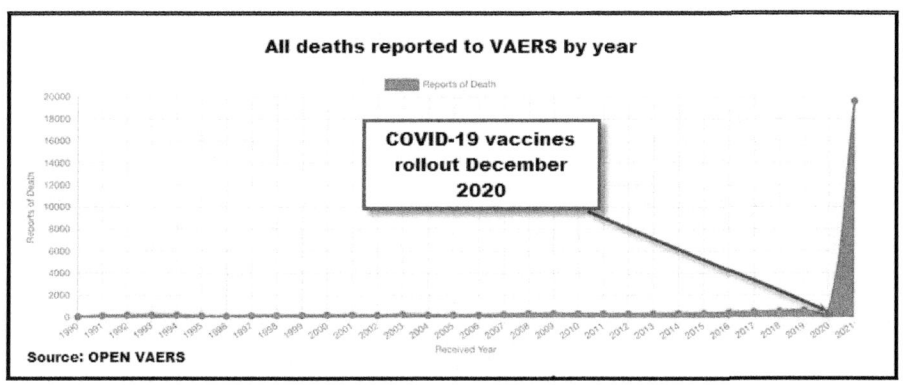

And they want to make these things compulsory?

People try to defend these numbers with the argument that deaths are inevitable when hundreds of millions of doses have been administered, but so what? Does that make it acceptable? The deaths are completely unnecessary when we have provably effective treatments like ivermectin and hydroxychloroquine available. If you drill down into the VAERS data you can see individual reports. Look at the ages of some of the deceased – right down to young children – and weep. Remember, 53 deaths from 45 million doses with swine flu in 1976 warranted the end of that programme. What has become of the world that this slaughter is allowed to continue?

The scary thing is that the VAERS system is known to suffer from chronic under-reporting. The real figure is, without a doubt, far higher and probably many multiples of 20,000 deaths. People suggesting that the spike in the graph 'doesn't prove causation' probably work in the business of disinformation. Either that or they might be vaccine safety data analysts for the CDC, but I repeat myself. Besides, this is not what VAERS was created for. VAERS is a reporting system for adverse events and deaths suffered after a treatment. These people received a COVID-19 vaccine and then they died. Period.

Do not be fooled by the FDA's much-heralded 'approval' of Pfizer's Comirnaty vaccine. Concurrent with the announcement on 23 August 2021 that Comirnaty had been approved, was the admission Comirnaty was unavailable due to insufficient stocks. We were also led to believe that the Comirnaty vaccine is interchangeable with the Pfizer BioNTech COVID-19 vaccine.

How can this possibly be true when one is approved and the other has Emergency Use Authorisation only? Legally they are not the same thing. As Meryl Nass wrote the day after this 'approval':[112i]

"EUA products are experimental under US law. Both the Nuremberg Code and federal regulations provide that no one can force a human being to participate in this experiment. Under 21 US Code Sec.360bbb-3(e)(1)(A)(ii)(III),[112ii] 'authorization for medical products for use in emergencies,' it is unlawful to deny someone a job or an education because they refuse to be an experimental subject. Instead, potential recipients have an absolute right to refuse EUA vaccines."

Under the EUA, Pfizer is shielded from liability. Guess which jab was available to Americans in seemingly unlimited supply and which was conveniently not available? It looks an awful lot as though people have been tricked into taking a vaccine they believe has been approved, but is, in fact, an experimental vaccine permitted for emergency use only and where the manufacturers get off scot-free if and when things inevitably go wrong.

With a wide range of therapeutics and prophylactics available for COVID-19, it is patently obvious that no vaccine is necessary for this disease. Not even for the individual (and that goes double for children), let alone entire populations and never mind the whole world.

That said, the world's media is obsessed with promoting vaccines and sadly a lot of the public have been persuaded. Could the Moderna vaccine become the leading vaccine of the future by default? If so the Bill Gates and Pentagon-linked company will be working with blockchain provider IBM to have it injected into as many arms as possible. Blockchain technology is put forward as a solution to so many world issues as if it is somehow indispensable to humanity, yet blockchain did not even exist a decade ago. One sees the word 'blockchain' and wonders if IBM said something similar in 1933 about punch cards?

It's probably no reason to be concerned. Only a real conspiracy theorist would suspect that the US military, big pharma and a bunch of Malthusians might have nefarious motives in wanting to vaccinate the whole world and use blockchain to manage the process.

Gates even said in April 2020: "I suspect the COVID-19 vaccine will become part of the routine newborn immunization schedule."[101]

Make no mistake, they are coming for the children with these vaccines. A plan to vaccinate newborns when over 99.999% of newborns survive COVID-19 cannot be for health reasons. There must be *some* reason though. So what is it?

Could it be to surveil us digitally and track everybody at all times? Why yes, it could!

Bill Gates has expressed his view on the record that the US needs national disease surveillance and a tracking system.[113]

Gates also suggested this could involve vaccine records embedded in our bodies using a quantum dot tattoo with invisible ink. This was described in fine detail in a Science Translational Medicine paper from December 2019, a study that Gates himself had funded.[114]

Forget microchips being injected by vaccine, a straw man argument that the 'Fact-checkers' love to 'debunk' as a conspiracy theory. People will be monitored by quantum dot laser tattoos and disease databases if Bill Gates has his way and it is hardly a conspiracy theory when they tell you about it.

To the people who label any alternative view as 'conspiracy theory' and the intrepid fact-checkers whose job it is to denigrate those daring to question the motives of the wonderful, amazing, philanthropic Bill Gates (and who are often paid by Bill Gates), consider this.

It is a simple statement of fact that the Rockefeller and Gates families and foundations have had intergenerational ties. The Gates family and IBM have also had connections going back decades and we know that IBM and Rockefellers both had very close links to the Nazis. These are just facts. It is hardly six degrees of separation.

Now, look at this picture of the Carnegie 'Medal of Philanthropy 2001' presentation in the context of Malthus, population control, eugenics, Nazi ideology, vaccines and philanthropy.

Is it getting dark enough to see the light?

2.30

None Dare Call It Eugenics after Hitler

It was no surprise that after World War II the eugenics movement lost momentum. The very word 'eugenics' still carries a terrible stigma and it is rightly agreed to be morally reprehensible quack pseudoscience. Today, nobody will publicly admit to being a eugenicist any more than they would admit to being a paedophile. This does not mean these people do not exist though. It has just become slightly more difficult to identify them (but not that difficult).

Hitler's ghastly deeds may have derailed the movement but they most certainly did not sound the death knell. It was merely rebranded with much of the work continuing in other guises. Abortion, family planning and euthanasia campaigns were well funded and still are. Of course, the propaganda about overpopulation never went away and it probably never will.

As medical technology advanced, a new form of eugenics came upon the scene: human genetic engineering. Human genes can be changed or removed to prevent or cure disease and in theory, they can be altered to improve the human body in some way. The potential health benefits of gene therapy are amazing. Who could argue against these noble aims?

Improving the human condition was a goal of eugenics all along. Don't take this as some sort of endorsement of the movement, but although the early 19th Century eugenics advocates involved murder, sterilisation and forced segregation, the stated goal was to create what they believed to be a superior species. Today, people discuss the rights and wrongs of designer babies, which is at least an improvement on talking about sterilising imbeciles or segregating Italian immigrants.

Some of the organisations that originally funded and popularised eugenics in the UK and the US, now have leading roles at the WHO, including the Wellcome Trust, the Rockefeller Foundation and the Sterilization League for Human Betterment (today renamed 'Engender Health').

One eugenics organisation that rebranded as a centre of excellence for human genetic study is now involved in the production of COVID-19 vaccines – the aforementioned Galton Institute with the AstraZeneca-Oxford vaccine.[115]

Human genetics, genome editing, CRISPR technology and gene sequencing have become the areas lavishly funded by the same kind of people and groups so interested in eugenics a century ago. Scientists have now unravelled the entire human genetic code.

Today, any individual can be biologically identified and classified by trait and ancestry in a few days at a cost of a few hundred dollars. Dedicated, honest, well-meaning scientists have done amazing work in these fields and there is no question that this work could be used to benefit humanity.

2.31

Science Fiction is Already Here

Some of the breakthroughs achieved in genetics and other areas of medical research seem almost like science fiction. For instance, nanotechnology can deliver drugs to the human body and even diagnose medical conditions, communicating with external devices and pre-emptively warning of potential health issues.[116]

We are being prepared for a future where these nanobots will be injected into us to communicate with health care providers. This is not a conspiracy theory; it is a very well-documented aspiration of health care providers, data suppliers and IT companies, who have laid out their proposals in white papers and business plans.

For many of us, doctors will eventually be replaced by artificial intelligence and nanotechnology. Big data and AI will power the health industry for the masses. Online consultations with doctors are not the same thing as face-to-face consultations – hardly a controversial statement. These unsatisfactory experiences will probably hasten the introduction of nanotechnology because, compared with Zoom consultations, nanobots won't seem like much worse of an option. People will simply say, "Why not?" almost in despair.

It all has this feeling of inevitability about it, like the cashless society and digital currencies will feel like an 'inevitable' progression. But these are not inevitable events. They were planned and engineered with great effort.

What about the field of 'bioelectronics'? The National Institute for Safety and Technology reassuringly told us in a 2009 white paper that this endeavour:

"... has the potential to significantly impact many areas important to the nation's economy and well-being, including healthcare and medicine, homeland security, forensics, and protecting the environment and the food supply."[117]

Would you like an electronic device inserted into your body to protect homeland security, the economy and the environment?

When people are required to have regular booster vaccinations to keep their vaccine passports in order, some might reason that they might as well have a nanobot or bioelectrical device inserted during the same visit. When people are used to receiving experimental vaccinations regularly, these treatments will begin to look the same.

You've probably heard about the 'Internet of Things', where appliances like fridges, televisions and radiators all communicate with each other (and of course to Big Brother). How about the 'Internet of Bodies'?

You had better believe this is real because our friends at the WEF are enthusiastically selling it to us. According to the WEF's website from June 2020:

"... it's now time for the Internet of Bodies. This means collecting our physical data via devices that can be implanted, swallowed or simply worn, generating huge amounts of health-related information."[118]

The WEF is obsessed with collecting your data, data being the new oil and all that.

The rationale for this data collection stems from the United Nations 17 Sustainable Development Goals (SDGs). An awful lot can be traced back to these 17 Goals. The reason the WEF and the UN believe they need this information is to better implement the 17 SDGs. Those who wish to know which direction the world is heading in would do well to familiarise themselves with these SDGs. They affect every living soul, which the UN never tires of reminding us with their mantra 'leave no one behind'. When one deciphers the inane, flowery language these institutions are so fond of and gets to the heart of the matter, these goals do not look so appealing. 'Leave no one behind' begins to sound like the utterings of a school shooter after a while.

The Internet of Bodies is not a new concept either. A Georgia Tech paper from back in 2015, speaks of 'a seamless interconnection between today's cyber-world and the biological environment.' It shows how humans will be implanted and joined up to the internet. Its title: *The Internet of BioNanoThings*.[119] The 'biological environment' means you and I all joined up to the cyber world and I can assure you I did not give anybody my consent for this.

Projects like 5G and Elon Musk's Starlink would help make these ambitions a reality, which is probably why anyone who dares to criticise 5G is viciously smeared and gaslighted, despite the enormous body of evidence that electromagnetic fields (not just 5G) can cause a range of negative health issues in biological life forms. A *Lancet* meta-study from December 2018, found that:

"A recent evaluation of 2,266 studies (including in-vitro and in-vivo studies in human, animal, and plant experimental systems and population studies) found that most studies (n=1546, 68·2%) have demonstrated significant biological or health effects associated with exposure to anthropogenic electromagnetic fields."[120]

This is ample grounds to at least question the safety of 5G but if you oppose an antenna being placed near your local school you are branded a kook.

How about this terrifying comment in a 2002 paper on bionanobots produced by the Arlington Institute which says, under 'Defense applications':[121]

"In unconventional terms, bionanobots might be designed that, when ingested from the air by humans, would assay DNA codes and self-destruct in an appropriate place (probably the brain) in those persons whose codes had been programmed."

Someone wrote that in a paper! A nanobot could innocently be inhaled by a target, perhaps spread through the air conditioning system, then recognise a particular person's DNA and get to work destroying that person's brain. The victim might suddenly collapse and die with an aneurism and nobody would know why. This was just speculation in 2002. What are the chances that this technology exists today?

Bill Gates funded a project with $20 million to create an implantable birth control microchip.[122] The technology enables a woman's fertility to be switched on or off remotely and is claimed to last for up to 16 years.[123]

Elon Musk's Neural Link is another extremely creepy project. You can see a Neural Link video demonstration presented by Elon Musk himself online.[124] A small piece of human skull is removed from the patient and an electronic implant is embedded which links up to the human brain and of course the internet. As Musk enthuses, it's just a small piece of skull that needs removing and you can hardly see the scar! The procedure is performed by a robot. How reassuring.

To show what the chip is capable of, Musk released another video that shows a brain-chipped monkey playing a computer game with just the power of his mind.[125]

Some of Klaus Schwab's comments are positively bone-chilling.

"What the fourth industrial revolution will lead to is a *fusion of our physical, digital and biological identity*," Schwab told the Chicago Council on Global Affairs in November 2020.[126] (Emphasis added.)

Klaus Schwab is straight-up telling us he wants to merge humans with machines. Who is the unelected Klaus Schwab to make these assertions about what will happen to our bodies? Klaus can do what he likes with his own body and perhaps he already has (which might explain his personality), but he needs to leave the rest of us the hell alone.

Klaus Schwab's terrible books contain further details of his wish list, including devices that are 'implantable in our bodies and brains'. These devices can 'manipulate our own genes, and those of our children.'[127]

Schwab even speaks of microchips that can 'intrude into the hitherto private space of our minds, reading our thoughts and influencing our behavior.'[128]

This mind-reading technology sounds fanciful but if it ever does become available the message from me is, "Sorry, Klaus, but those four square inches are mine." That is a red line that will never be crossed and if it does, it will be over my cold, dead body. These technocrats' appetite for surveillance appears to know no bounds.

Schwab asks, "Where do we draw the line between human and machine?"[129]

This is where Klaus Schwab outs himself as a transhumanist. Thankfully, these people are far and few between but their small clique – several of whom reside in Silicon Valley – is an immensely rich and powerful one. Anybody who genuinely needs to ask this question is probably not to be trusted in a position of power, yet the entire WEF dances to the tune of this absolute nutcase. We could go on and on.

We should think twice before we allow corporations to insert electronic devices into our bodies. Remember that they can communicate with other devices and even be controlled remotely, all connected to the internet. This is something to consider in a business climate where the mantra is 'data is the new oil' and companies scavenge for every last data point.

Likewise, we should exercise the utmost caution with experimental vaccines that can programme our RNA. The matter-of-fact description of Moderna's technology on its website is informative.

"… we set out to create an mRNA technology platform that functions very much like an operating system on a computer."

The spiel continues:

"It is designed so that it can plug and play interchangeably with different programs. In our case, the 'program' or 'app' is our mRNA drug."[130]

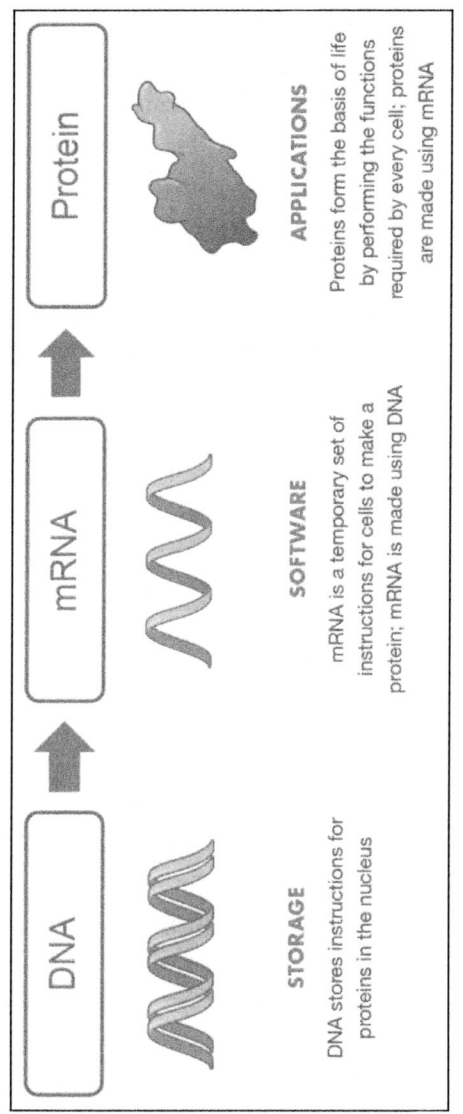

One wonders if the boffins at Moderna might consider a PR company to review their communications next time. The 'computer' in this case is the human body. That is essentially how Moderna's scientists view human beings – something to be programmed.

Chips in the brain are not a conspiracy theory and are being showcased personally by Elon Musk. Klaus Schwab admits he wants to fuse our physical, digital and biological identities. By allowing corporations to insert devices and novel vaccine technologies into us we are allowing corporations to get their hands on our innards.

There's pretty much nobody I would trust with any of this technology, least of all people who associate with and fund convicted paedophiles (Bill Gates with Jeffrey Epstein), corporations that have been fined billions for dishonesty and medical malfeasance (most of the pharmaceutical industry) or Pentagon-funded companies that boast about turning human bodies into operating systems (Moderna), let alone the creepy Elon Musk, who I truly believe is insane. Chip your brain if you want to Elon, but leave the rest of us alone, please.

If it could be demonstrated that *any* of this equipment is necessary then it might be entertained as the progress they want us to believe it is. But none of it is. None of it.

If a device was able to predict an incipient cardiac arrest and a team of medics dispatched to your house just in time to save you, that would be fine. But when a person sheds a pound or two, it is nobody's business but their own that they just went to the toilet. I would rather anonymous data analysts did not know my blood pressure increases when I listen to ghastly electronic music and I certainly don't want some erstwhile healthcare employee to know that my heart rate speeds up when that particular girl walks into the room. Of course, they will be monitoring her location too and everything will be linked up, with the AI noticing these irregularities. The computer and anyone with access to this data will know that I fancy her even if she does not.

Think how information like that might be used for malicious ends. From your vital observations, they will know which chocolate bar you like best, what music makes you happy and how much you can't stand the sight of Piers Morgan on a scale of nine to ten. Think how the media already demoralises the public with climate change and COVID-19 doom and how they practically troll the public with stories about Donald Trump. Imagine if propaganda could be individually tailored to people based on their prejudices, fears and actual physiological responses. An AI-driven algorithm might end up being the *cause* of the heart attack in a victim that the AI would then be required to dispatch medics to attend to.

The point is that this information is none of anybody's damned business. Understand that this data will all be swept up and ultimately be used to further the sustainable Development Goals, for the 'greater good' or whatever reason their sponsors tell us it is for. Do you trust the UN and WEF to have your best interests at heart? Goal 2 is 'Zero Hunger' and we've seen how the UN has failed so abjectly to achieve this in the past. (Might I suggest they have 'zero interest' in 'zero hunger' and are more interested in 'zero population growth'?)

These devices are always promoted with health benefits in mind and of course, they could be used for good purposes. I'd be thrilled if I believed the people funding all of this cared about my health or wanted a better world for us. Sadly the evidence suggests that this is not the case.

Decisions decisions...

To wit, look at the proportion of grant money spent on studying dietary interventions. If even 1% of the amount spent on genetic research was diverted to studying nutrition, then most diseases would be under control by now, especially cancer, heart disease and diabetes, all of which can be reduced and even reversed by simple dietary interventions. Yet these diseases are claiming increasingly large numbers of victims. Successful treatments for these diseases are routinely hushed up[131] and outright suppressed by the industry and their regulatory bodies, just as they are with COVID-19.

The fact that nutrition is almost ignored by the rich foundations and philanthropists is all the evidence one needs to understand that they do not care about our health. But they do care about genetics. They really, really do care about genetics.

Whoever is in charge of appointments at the White House these days (obviously not Joe Biden) has promoted Eric Lander, the Massachusetts Institute of Technology (MIT) professor and geneticist who co-led the Human Genome Project, to the position of top scientific adviser to the President. Eric Lander's appointment received some unfavourable media attention due to his relationship with Jeffrey Epstein.

Before he totally killed himself in a prison cell just after the cameras in the cell stopped working, serial paedophile and fixer Jeffrey Epstein boasted on his website that he had funded Eric Lander with grants for scientific work, as well as many other scientists.[132]

"The Foundation has also had the priviledge [sic] of sponsoring many prominent scientists including Gerald Edelman, Murray Gell-Mann, Eric Lander, George Church, Stephen Hawking, Kip Thorne, Marvin Minsky, Lawrence Krauss, Lee Smolin and Gregory Benford and Nobel laureates, Gerard 't Hooft, David Gross and Frank Wilczek."

Lander's spokesman denied this was true, understandably, because who would want to be associated with Jeffrey Epstein?[133]

The denial is questionable because nearly all of Epstein's other boasts about funding scientists turned out to be true. Why lie about this particular donation? Epstein's ties with Eric Lander's employer, the MIT on the other hand, cannot be denied. The paedophile had funded the university's MIT Media Lab with millions of dollars, some of it apparently on behalf of Bill Gates according to *The Guardian*. All the while, the recipients of this money were fully aware of Epstein's sordid, child molesting lifestyle.[134]

Eric Lander's denial was only made after Epstein was arrested in 2019, but Epstein had already been convicted for child sex offences back in 2008. Lander had certainly met Epstein, as can be seen in the 2012 photograph below taken from Epstein's website.[135] One supposes that online 'fact-checkers' could try to claim that Epstein photo shopped the image and posted it to his website (you know, just for kicks).

Other photographs of the two men from Epstein's archived website have thumbnails that read 'Eric Lander describes genetic sequencing with Jeffrey Epstein' and 'Jeffrey Epstein and Eric Lander discuss genomes'.

Eric Lander relaxes with Jeffrey Epstein, photo from the archived website of Jeffrey Epstein.

Apparently, one of the projects that the hideous child trafficker Jeffrey Epstein was fascinated in, was summed up in this *New York Times* headline:[136]

'Jeffrey Epstein Hoped to Seed Human Race With His DNA.'

His plan was some bizarre artificial intelligence/transhumanist/genomic sequencing project that involved Epstein impregnating dozens of women who would live on one of his ranches. The sickness of Epstein is mind-boggling. It is hard to fathom how a man who spent his days generating blackmail material against politicians and celebrities in between having sex with children could consider that the world needed *more* of his type. That is some next-level narcissism, hardly congruous with the mentality of a man who tops himself.

This is not to suggest Eric Lander had anything to do with these vile plans. Nor can this author produce documentary evidence that Lander did take money from Epstein, although those denials look flimsy. It is certainly not to Lander's credit that he would converse genially with one of the world's most despicable sex offenders about genetic sequencing. When this picture was taken, Epstein was a convicted child sex offender and Eric Lander had to know that. Is this an appropriate person to advise the President?

In January 2016, Michael Eisen, Berkeley professor and the editor of the journal *eLife*, said on his blog about Eric Lander that:[137]

"There is something mesmerizing about an evil genius at the height of their craft, and Eric Lander is an evil genius at the height of his craft. Lander's recent essay in CELL entitled *The Heroes of CRISPR* is his masterwork, at once so evil and yet so brilliant that I find it hard not to stand in awe even as I picture him cackling loudly in his Kendall Square lair, giant laser weapon behind him poised to destroy Berkeley if we don't hand over our patents."

As Buzzfeed reported:[138]

"Under Biden, genome-sequencing pioneer Eric Lander may become the most powerful scientist in US history."

Eric Lander's role will be elevated to a cabinet position for the first time.[139] This is the man who held meetings with the world's most notorious paedophile after he had been convicted. He took money from him – allegedly – and his MIT colleagues had even better relations with Epstein and took even more money from him for a fact. Eric Lander is a person of interest when it comes to the American government's future investment in scientific research, in particular genomic research. It would be good to know who really decided to put Eric Lander in this position, and why.

Jeffrey Epstein, like Bill Gates, called himself a 'philanthropist'. Virtually every last one of these so-called philanthropist billionaires maintains the world is overpopulated. We know that is a lie so why should we trust them when it comes to our health?

Rather than promoting good health, might we suggest that the technologies we have mentioned are going to be used as surveillance tools so the human cattle can be micromanaged? Things will be bad enough if the anonymity of cash is lost and every last one of our digital currency-based transactions are tracked and traced. How bad will things be when our bodies are also tracked?

All this data will be transmitted wirelessly across the internet meaning it can be intercepted. These systems can be hacked and they no doubt will be. Criminal hackers are probably less of a worry than the people who are supposed to be trusted with the data though. The people who set these systems up did so for the precise purpose of surveilling and micromanaging our lives. They have big plans for all this data. Hackers only want to earn a bit of illicit cash.

Those who believe that the data is anonymised and only used for relevant purposes and blah blah, are sadly mistaken. Let's not fall for the notion that this is only about money either. There is more to this than greedy companies targeting us with personalised solutions.

This total loss of privacy and feeling that your every move is being recorded, even inside your own home, is corrosive to the human spirit. People already self-censor when it comes to expressing opinions in case they are marked out or their name put on a 'list'. Imagine forgoing a small pleasure like a chocolate bar just to keep the AI 'happy', or trying not to have thoughts about personal freedom in case your mind was being read by your brain chip.

If all this technology and knowledge were to be placed in the wrong hands there are huge ethical questions to be asked at the very minimum. Now look at whose hands it is actually placed in.

Doubtless, the vast majority of scientists involved in modern human genetics projects are thoroughly decent people with good intentions. One cannot be so confident saying the same thing about the leaders of the organisations funding their work – for instance, the Rockefeller Foundation, the Bill & Melinda Gates Foundation and the Wellcome Trust.

If this sounds a tad cynical, consider this question: with all these incredible breakthroughs in science, why do we witness simultaneous explosions in the rates of heart disease, diabetes, autism, Alzheimer's, dementia, cancer and plummeting fertility (*cough* glyphosate)? Life expectancy may have increased over the last century, but people are sicker and prescribed more drugs.

Asking a similar question to the one we raised about the UN's attempts to solve world hunger, do the people who fund all this amazing work want us to be healthier or not? If they do, their efforts have been an abject failure. If they do not then what are the real goals of these programmes?

In attempting to answer that question, it is almost impossible to escape the fact the suppliers of the technology all believe in depopulation.

2.32

From Eugenics to Straight-Up Depopulation

The popularity of eugenics might have been waning since World War II but, as the world's population grew from 2.5 billion to almost eight billion, the calls for depopulation grew louder and louder. They came from all angles.

Bertrand Russell was a British aristocrat and a serious intellectual force. In 1953 he wrote:[140]

"I do not pretend that birth control is the only way in which population can be kept from increasing. There are others, which, one must suppose, opponents of birth control would prefer. War, as I remarked a moment ago, has hitherto been disappointing in this respect, but perhaps bacteriological war may prove more effective. If a Black Death could be spread throughout the world once in every generation survivors could procreate freely without making the world too full... The state of affairs might be somewhat unpleasant, but what of that? Really high-minded people are indifferent to happiness, especially other people's."

It is somewhat ironic that Bertrand Russell was a pacifist in World War I – and good on him for that – yet had no objection to millions dying in wars or plagues, being 'really high-minded' and all that.

Bertrand Russell considered himself a liberal and a socialist. He was in fact, a propagandist and a subversive. He was an honorary chairman of the Congress for Cultural Freedom (CCF), an institution whose output would have influenced many a useful idiot in academia or politics unaware that the CCF was a project funded by the CIA to nudge the political left towards the right. Compromised authors and journalists who wrote for the CCF came in very useful in the wake of the JFK assassination when legitimate doubters were mocked and maligned as 'conspiracy theorists' by CCF stooges.

Bertrand Russell gave intellectual cover for evil people and, like so many of his type, he supported a world government. Naturally, such work does not go unrewarded. Bertrand Russell was a Nobel laureate (for literature, 1950) and, as is almost mandatory for scientists who voice depopulation beliefs, he was elected as a Fellow of the Royal Society.

A host of ghouls would step up to carry the baton for world depopulation. As usual, all of them would be utterly wrong but many would spin fine careers out of their humanity-hating diatribes, just as

Malthus, Galton and George Bernard Shaw had done in the 18th, 19th and 20th Centuries.

Take Stanford professor Paul Ehrlich. In his 1968 book *The Population Bomb* he wrote:[141]

"A cancer is an uncontrolled multiplication of cells; the population explosion is an uncontrolled multiplication of people... We must shift our efforts from the treatment of the symptoms to the cutting out of the cancer. The operation will demand many apparently brutal and heartless decisions."

'Brutal and heartless decisions' to cut out the human 'cancer'. That is the kind of language that will get a man recognised in certain circles. The very first sentence of his book set the tone. With world population at around 3.5 billion he wrote:

"The battle to feed all of humanity is over. In the 1970s, the world will undergo famines – hundreds of millions of people will starve to death in spite of any crash programs embarked upon now."[142]

He warned of theoretical doomsday scenarios that might occur if we were not careful, although one cannot be sure if these are not his fantasies. The following quotes are taken from the 1970 edition of the book.

"Only the outbreak of a particularly virulent strain of bubonic plague killing 65% of the starving Egyptian population had averted a direct Soviet-American clash in the Mediterranean."

"The third Los Angeles killer smog in two years has wiped out 90,000 people... The President's Environmental Advisory Board has reported a measurable rise in the sea level due to melting polar ice caps. [The Board] recommends the immediate compulsory restriction of births to one per couple, and *compulsory sterilization of all persons with I Q scores under 90*." ([Emphasis added.)

It if looks a bit like eugenics, that is because population control always is. Ehrlich even wrote about a scenario where the Pope would condone abortion one day.

"Pope Pius XIII, yielding to pressure from enlightened Catholics, announces that all good Catholics have a responsibility to drastically restrict their productive activities. He gives his blessing to abortion and all methods of contraception."

Ehrlich would make a slew of other ludicrous predictions. He claimed that India wouldn't be able to handle its population growing from 400 million to 600 million because they would never be able to feed themselves. The 1.4 billion population living in India today would beg to differ. Rates of malnutrition and starvation are far lower today than they were when the population stood at 400 million and 600 million.

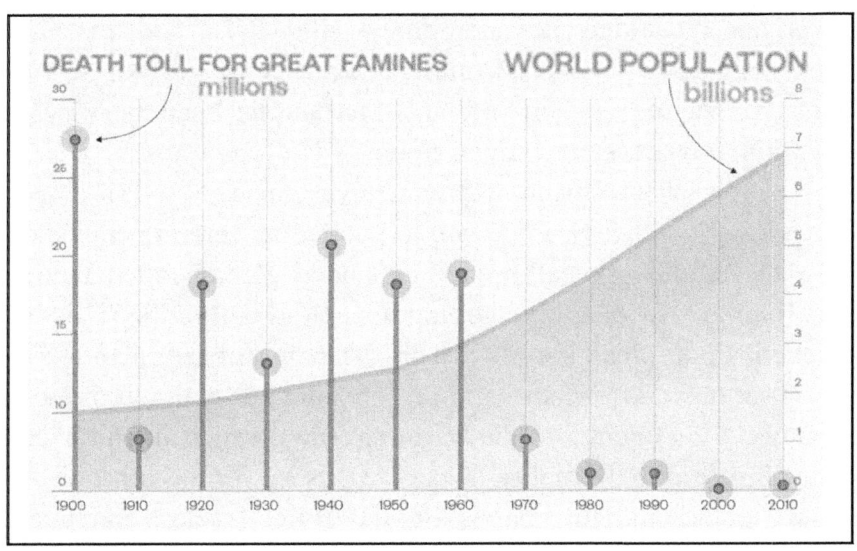

Ehrlich – wrong again.

Ehrlich also said in 1971, "If I were a gambler, I would take even money that England will not exist in the year 2000."[143]

None of his hare-brained predictions harmed Paul Ehrlich's academic career and of course, England still exists. Ehrlich was even elected as a Fellow of The Royal Society in 2012, just to show there were no hard feelings felt by his English counterparts.

Instead of *The Population Bomb*, Ehrlich might as well have called his book *Bomb The Population*. He probably would have enjoyed just as successful a career. *The Population Bomb* sold over two million copies and genuinely did influence policymakers.

Barack Obama's chief science czar, John Holdren, co-wrote a 1,051-page book with Paul Ehrlich and his wife, Anne Ehrlich, in 1977 called *Ecoscience* where they advocated a policy of sterilisation by drugging the water supply.[144] They also suggested forced sterilisations and abortions and mandatory body implants to prevent women from having children.[145]

They spoke of a powerful 'planetary regime' that would use a global police force to achieve this.

"Perhaps those agencies, combined with UNEP and the United Nations population agencies, might eventually be developed into a Planetary Regime – sort of an international superagency for population, resources, and environment."[146]

Have you ever noticed that whenever there is a crisis, be it financial, political, environmental, or pandemic, certain people come out of the woodwork calling for enhanced global governance? In John Holdren's case, he was getting his call in before the crisis even happened. Such pleas are always in the name of co-operation and harmonisation, for what kind of person doesn't approve of international co-operation and harmonisation, especially in a crisis (real or imagined)? In reality, these are pushes for the centralisation of more power into the hands of malleable, faceless bureaucrats appointed by billionaires, powers that just happen to weaken nation-states and their governments.

Can you possibly guess which non-toxic, non-polluting atmospheric gas John Holdren's concern about the environment stemmed from?

"As University of California physicist John Holdren has said, it is possible that carbon dioxide, climate-induced famines could kill as many as a billion people before the year 2020."[147]

Of course, it was the evil carbon dioxide that every living creature needs to survive. The 20 parts per million of manmade CO_2 in our atmosphere did not cause a billion deaths by famine you will be amazed to learn, but John Holdren was well aware of which panic button to press.

John Holdren is the kind of man who gets to help shape national policy, also serving as one of President Bill Clinton's science advisors from 1994-2001. Pointing this out is not to pick on Bill Clinton or Barack Obama for their bad taste, or even the Democrats, although it is interesting to note yet again that the so-called progressives contain some seriously unsavoury people in their ranks. Not all of these high-level advisors are personally handpicked by the President of the day. It is most likely that several key advisers are allocated to the President by the people who control the real levers of power. High-value appointments are sometimes retained whoever is President, Dr Anthony Fauci being a prime example. He has been in situ since 1984 as the head of the National Institute of Allergy and Infectious Diseases (NIAID). Interestingly, Wikipedia says that Fauci has been offered the position of director of the NIH several times, but has declined each time.[148]

It would appear that somebody important wants the crafty little gnome exactly where he is.

2.33

Mr Fixer

"Who controls the food supply controls the people; who controls the energy can control whole continents; who controls money can control the world."

Henry Kissinger

Probably the most famous Presidential adviser since World War II has been Henry Kissinger, a man who has had the ear of every US President for over five decades. Kissinger was a protégé of Nelson Rockefeller and it is an open secret that it is not the US government

Kissinger has been running errands for all these years, let alone the American people, but the interests of the Rockefellers.

In 1974, a year after the mass terror bombing of Cambodia which he had orchestrated, Henry Kissinger authored a policy document with the bland title *National Security Study Memorandum 200*.[149] The document was obsessive about population growth in developing countries and the need for America to reduce this growth. Kissinger targeted several countries whose populations, in his view, needed curtailing lest they gain economic, political and military strength.

"The US economy will require large and increasing amounts of minerals from abroad, especially from less developed countries... That fact gives the US enhanced interest in the political, economic, and social stability of the supplying countries. Wherever a lessening of population pressures through reduced birth rates can increase the prospects for such stability, population policy becomes relevant to resource supplies and the economic interests of the United States..."

In other words, prevent population growth in case these countries consume their own natural resources. 'Economic interests of the US' is a more honest appraisal of their grasping intention than 'National Security'. The naked greed of this policy was illustrated in the very same section of the document, where it was admitted: "The US, with 6% of the world's population, consumes about a third of its resources."

The policy was long on greed but short on patience.

"We cannot wait for overall modernization and development to produce lower fertility rates naturally since this will undoubtedly take many decades in most developing countries."

Kissinger's *National Security Study Memorandum 200* was only declassified in 1989, but history shows that the document has served as a blueprint for US foreign policy to this day. Henry Kissinger was awarded a Nobel Peace prize in 1973.

These objectives would be tackled with tools such as food scarcity, sterilisation and war. It is hard to know just how many millions of innocents have died or not been born as a result of

Kissinger's lust for power and domination. It certainly makes one think about those declines in fertility rates we looked at earlier.

When lamenting the UN's hopeless track record for reducing famine, it probably did not help matters that its most influential member state, the USA, has been actively trying to curb population growth with a foreign policy resembling genocide.

As is often the case, lethal foreign intervention would fall under the foreign "aid" budget.

The US Agency for International Development (US-AID) is a federally funded organisation with a long track record of providing cover for the CIA.[150]

One imagines the citizens of the third world must pray that the US and UN would stop 'helping' them so much. Kissinger's document continued:

"Implementing the actions discussed above... will require a significant expansion in AID funds for population/family planning. A number of major actions in the area of creating conditions for fertility decline can be funded from resources available to the sectors in question (eg education, agriculture). Other actions, including family planning services, research and *experimental activities on factors affecting fertility*, come under population funds." (Emphasis added.)

Promoting fertility decline via the agriculture budget? How does that work? (Have a serious think about that question.) And what on earth are these 'experimental activities on factors affecting fertility'?

They couldn't possibly be talking about vaccines could they? One is reminded of Bill Gates ominous remark from that 2010 TED Talk.[37]

"The world today has 6.8 billion people. That's headed up to about nine billion. Now, if we do a really great job on new vaccines, health care, reproductive health services, we could lower that by, perhaps, 10 or 15 percent."

I would just like to know what Bill Gates meant by that statement other than exactly what is written. Is it wild speculation or paranoia to imagine that vaccines might be used to slow down population growth? There have been many unfortunate incidents that lend weight to this suspicion, so let's take a look at a few of them.

(Quick message to would-be detractors who have a hard time believing the following information: save yourself the bother of fact-checking any of the following examples with online 'independent fact-checkers'. Virtually all of these websites are vehicles of disinformation, often personally funded by the likes of Bill Gates, George Soros and Pierre Omidyar, along with Silicon Valley billionaires and NGOs that push the depopulation agenda. Try to use real sources instead and, if having done so you still maintain that any of the following is fake news', you are welcome to take legal action.)

2.34

Tetanus Vaccines in Kenya

In 2014, doctors from the Catholic Health Commission of Kenya grew suspicious that only girls and women between the ages of 14 and 49 were being targeted with the tetanus vaccine, which was being provided with assistance from the WHO and UNICEF. They released a statement questioning why the campaign left out young girls, boys and men even though they were also vulnerable to tetanus and asked, "In the midst of so many life-threatening diseases in Kenya, why has tetanus been prioritized?"[151]

They discovered that a sample of vaccines that had been administered to 2.3 million females had been contaminated with the anti-fertility hormone Human Chorionic Gonadotropin (hCG). Dr Muhame Ngare of the Mercy Medical Centre in Nairobi said, "We sent six samples from around Kenya to laboratories in South Africa. They tested positive for the HCG antigen... They were all laced with HCG." Dr Ngare also stated, "This evidence was presented to the

Ministry of Health before the third round of immunization but was ignored."[152]

The Catholic Health Commission of Kenya statement declared, "This [vaccine] has previously been used by the same partners in Philippines, Nicaragua and Mexico to vaccinate women against future pregnancy."[151]

The WHO, UNICEF and Kenyan government all denied there was a problem with the vaccine. In 1995, when similar concerns were raised by the other countries mentioned, the WHO issued a rebuttal insisting there was no hCG in the tetanus vaccines they had supplied.[153]

But, laboratory testing showed that there was, not only in the Kenyan samples in 2014 but also in Mexico, the Philippines and Nicaragua in the mid-1990s. This was revealed in a *Pubmed* article from 1995, which said:[154]

"In the fall of 1994, the Pro-Life Committee of Mexico was suspicious of the protocols for the tetanus toxoid campaign because they excluded all males and children and called for multiple injections of the vaccine in only women of reproductive age. Yet, one injection provides protection for at least 10 years. The Committee had vials of the tetanus vaccine analyzed for hCG. ... *Similar tetanus vaccines laced with hCG have been uncovered in the Philippines and Nicaragua.*" (Emphasis added.)

Of course, the question remains: why only vaccinate females?

Dr Ngare said on behalf of the Kenya Catholic Doctors Association: "Either we are lying or the government is lying. But ask yourself, what reason do the Catholic doctors have for lying?"[152]

In September 2017, Kenya's opposition leader, Raila Odinga, estimated that 500,000 young girls and women may be infertile as a result of this tetanus vaccine programme.[155]

To compound this tragedy, there was no outbreak of tetanus in Kenya, only the perceived threat of tetanus due to local flood conditions.

2.35

Disaster 'Relief'

It appears to be a policy of the WHO and UNICEF to roll out mass vaccination programmes in areas where there have been local disasters. It might sound rational to target areas at risk of disease outbreak but recent examples show that this can just make matters worse.

In 2013, UNICEF ran a vaccination campaign in the Philippines after a typhoon had struck, with hundreds of thousands of doses of live oral polio vaccine and measles vaccines targeted at children. This was despite measles rates being minuscule in the Philippines (and falling) and also the fact children had already been vaccinated for measles previously. Was this an admission that the measles vaccine does not work? What is even worse is that there had not been a single case of polio in the Philippines since 1993.[156]

In the same year of 2013, a mass vaccination campaign with the live oral polio vaccine occurred among Syrian refugees. Syria had been reportedly polio-free since 1999 but when 13 new cases were reported UNICEF went to work. After the mass vaccination program started, cases of polio began to reappear in Syria.[157]

The oral polio vaccine is well known for causing the very disease it is supposed to prevent, with some victims suffering paralysis. It has been well documented in the scientific literature that the oral polio vaccine contains 'pathogenic vaccine-derived polioviruses'. [158]

For all the great publicity this vaccine receives, the fact is that people who have received the live vaccine can shed the poliovirus, which can end up in sewage systems causing outbreaks of polio and paralysis. The live oral polio vaccine has been discontinued in most of the world with an inactivated virus used instead. This requires an injection, however, and it is much easier to administer a live oral polio vaccine when mass vaccination is desired.

An editorial appeared in the *Oxford Journals Clinical Infectious Diseases* periodical as far back as 2005 titled: *When Can We Stop Using Oral Poliovirus Vaccine?*[159]

Yet the oral polio vaccine is still being used today with devastating effects.

If anybody is wondering where the 13 cases of polio that suddenly arrived in Syria came from in a country that had been polio-free since 1999, it is possible that the source was from the influx of NATO-backed foreign terrorists sent to dismantle Syria. Then again, polio-stricken men do not make the best fighters.

An alternative hypothesis is that it was caused by the vaccine itself. This is possible, if not the probable cause. It transpires that UNICEF had already been in Syria, from November 2012, vaccinating children under the age of five with the live oral polio vaccine.[160] It is perfectly plausible that UNICEF created the very outbreak they subsequently attempted to arrest.

In 2019, there was another outbreak of polio in the Philippines and even the WHO admitted that it was caused by the polio vaccines. *The Associated Press* reported:

"WHO and UNICEF said in a joint statement the polio outbreak in the Philippines is concerning because it is caused by vaccine-derived poliovirus type 2."[161]

The fact is that the vast majority of polio cases in the modern world are caused by polio vaccines themselves. It is inconceivable that UNICEF and the WHO are unaware of the scientific literature or the fact that most countries have discontinued using the oral polio vaccine. But their response to the outbreak in the Philippines, one they admit was caused by the vaccines, is simply to prescribe more vaccines.

"As long as one single child remains infected, children across the country and even beyond are at risk of contracting polio," UNICEF Philippines representative, Oyun Dendevnorov, said.[161]

Sounds like a great business opportunity if one is heavily invested in vaccine companies and holds great sway with international health organisations. As the Associated Press reported:

"At least 95% of children under age 5 need to be vaccinated to halt the spread of polio in the Philippines, WHO and UNICEF said."[161]

Claims such as '95% of children under age 5 need to be vaccinated to halt the spread' are pure conjecture. They have never been proven, just as the estimates for so-called vaccine 'herd immunity' with COVID-19 jabs have never been proven. The numbers are simply plucked out of thin air, whether the figure is 68%, 80%, 90% or 95%. Any person who makes such a claim is either ignorant or lying.

In any case, how can the WHO justify using the oral polio vaccine *at all* when they know what damage it is capable of causing? It seems quite apparent that UNICEF and WHO use these local disasters to move in and commence their mass vaccination exercises. These projects can increase local misery but they will always benefit the vaccine manufacturers. And, of course, the depopulation advocates.

2.36

The Discredited DTP Vaccine

There is another infamous vaccine that likely causes more harm than it prevents – the DTP shot, used to inoculate against Diphtheria, Tetanus and Pertussis. Its use has largely been discontinued in western countries but it is still used by UNICEF in poorer countries, again with deadly consequences.

In 2017, the Informed Consent Action Network (ICAN) issued a notice to UNICEF complaining in the strongest possible terms:[162]

"World-renown vaccine experts at the University of Southern Denmark, in one of the first-ever properly controlled comparisons of vaccinated children to completely unvaccinated children, concluded that children vaccinated with DTP were 10 times more likely to die in the first 6 months of life than the unvaccinated. UNICEF supplies first-world countries with what is considered a safer form of this vaccine, DTaP, but supplies to most third-world countries DTP. We demand that UNICEF cease distributing DTP vaccine or provide scientific support its DTP campaigns are not resulting in mass murder."

ICAN further pointed out: "This study also found that children vaccinated with DTP were dying from causes never associated with this vaccine, such as respiratory infections, diarrhea, and malaria. This indicated that while DTP reduced the incidence of diphtheria, tetanus, and pertussis, it increased susceptibility to other infections."[163]

People who support the pharmaceutical industry in these matters tend to claim side effects are always 'coincidental'. This is despite the fact these very effects are usually listed as potential adverse reactions in the manufacturers' own vaccine inserts.

The study's[164] lead author is the highly respected Dr Peter Aaby, with over 300 published studies bearing his name.[165]

Dr Aaby concluded: "All currently available evidence suggests that DTP vaccine may kill more children from other causes than it saves from diphtheria, tetanus or pertussis."[163]

The respected Dr Aaby's protests that the DTP vaccine was causing needless deaths were not addressed in UNICEF's bland response to ICAN. They completely ignored Dr Aaby's findings. Until UNICEF addresses the questions asked and provides 'scientific support its DTP campaigns are not resulting in mass murder' is it unreasonable to call it mass murder?

Vaccine proponents often claim that, where side effects arise from a vaccine, the risks are acceptable because of the overall good that they do. These cases illustrate a different story entirely. In many cases, the vaccination programmes are *causing* the diseases.

2.37

Israel's Ethiopian Policy

Evidence suggesting race-based sterilisation performed by the Israeli ministry of health was uncovered by an investigative journalist curious about the falling birth rate among Ethiopian immigrants.[166]

Under Israel's Law of Return, more than 100,000 Ethiopian Jews have travelled to Israel since the 1980s. However, investigative reporter, Gal Gabbay, discovered that while being held in immigration centres in Ethiopia, women were told they were required to be vaccinated as a condition of entry.

Former Israeli Prime Minister, Benjamin Netanyahu, warned that illegal immigrants from Africa 'threaten our existence as a Jewish and democratic state'.[167] As well as being the former Prime Minister, Netanyahu was also the Israeli Health Minister from 2009 to 2013.

Initially, it was denied that such a policy was being operated but, in February 2013, a tacit admission was made when the Israeli Health Ministry issued a statement with new guidelines forbidding the renewal of prescriptions of Depo-Provera 'for women of Ethiopian – or any other – origin, if there is the slightest doubt that they have not understood the implications of the treatment.'[168]

It will not be lost on readers that sterilising women on the basis of their race or religion is the same disgusting policy enacted in early 20th Century America and Germany in the 1940s, often with Jews as the victims.

2.38

USAID in Peru in the 1990s

A programme carried out in Peru in the mid-1990s, under the Presidency of Alberto Fujimori, led to the forced sterilisation of over 200,000 women. The National Catholic Register noted a helping hand from the US, so often the case in Latin American politics.[169]

"An unsettling aspect of the entire Peruvian campaign is the involvement of the US government. The specific agencies that were involved in Peru's sterilization campaign were the US Agency for International Development (USAID), the United Nations Population Fund (UNFPA) and the NIPPON Foundation (a Japanese nonprofit). It is known that UNFPA donated $10 million for the forced-sterilization campaign."

This rather looks like Kissinger's policy at work, namely reducing fertility, spelt out in Memo 200, carried out with US taxpayers' money with the assistance of the very agency he named, AID. Also, note the UN's not so benevolent assistance in this sterilisation spree.

As author Ryan McMaken notes:[170] "Consent from women apparently doesn't trouble American policymakers when it comes to funding and supporting population control policies in foreign countries."

Or even in its own country. In 1976, the US government admitted to having sterilised American Indian women without their permission, including women under the age of 21 despite an existing court order forbidding this very thing. The US General Accounting Office found that 3,406 American Indian women had been sterilised between 1973 and 1976.[171]

It always helps if a willing partner can be found in the target country. McMaken added: "USAID workers may have sensed an opportunity to partner with the Peruvian regime – which itself viewed the impoverished natives in the Andean highlands as 'problematic' – in efforts to implement a eugenics program in Peru".

The US government and its agencies have shown a particular skill at identifying, cosying up to and even installing such willing partners in countries it has economic interests in.

Many struggle to believe that people can be so evil. They would rather make excuses that these incidents must have all been accidents and that UNICEF was only trying to help. The trouble with that wishful thinking is that the results of the interventions chime perfectly with the stated aims of written policies: population reduction.

What is worse is that some people accept that these campaigns happen but are prepared to turn a blind eye because they believe the world is overpopulated. We have seen that this is the biggest lie of all, a lie designed to elicit exactly this response from the indifferent masses.

2.39

That's Not All Folks – Kennedy Speaks

In April 2020, Robert F Kennedy Jr, a qualified lawyer, called out Bill Gates very publicly with a list of claims that made very interesting reading. Let's look at some of these allegations and bear in mind the selected quotes are only about half of what RFK Jr wrote. [172]

"In 2010, the Gates Foundation funded a trial of GSK's experimental malaria vaccine, killing 151 African infants and causing serious adverse effects including paralysis, seizure, and febrile convulsions to 1,048 of the 5,049 children.

During Gates 2002 MenAfriVac Campaign in Sub-Saharan Africa, Gates operatives forcibly vaccinated thousands of African children against meningitis. Between 50-500 children developed paralysis. South African newspapers complained, 'We are guinea pigs for drug makers'.

In 2014, the Gates Foundation funded tests of experimental HPV vaccines, developed by GSK and Merck, on 23,000 young girls in remote Indian provinces. Approximately 1,200 suffered severe side effects, including autoimmune and fertility disorders. Seven died. Indian government investigations charged that Gates funded researchers committed pervasive ethical violations: pressuring vulnerable village girls into the trial, bullying parents, forging consent forms, and refusing medical care to the injured girls."

In 2017, the World Health Organization reluctantly admitted that the global polio explosion is predominantly vaccine strain, meaning it is coming from Gates' Vaccine Program. The most frightening epidemics in Congo, the Philippines, and Afghanistan are all linked to Gates' vaccines. By 2018, ¾ of global polio cases were from Gates' vaccines."

Social media companies can censor RFK Jr by closing his accounts[173] and the media can malign him with personal attacks. But for all the abuse and ad hominem attacks, it is rare to see RFK's claims disproven. To date, there does not appear to be any legal action arising from these claims and we know that Bill Gates can be litigious when he wants to.

Bill Gates has used his billions to influence the WHO and UNICEF, often to mass vaccinate people in poor countries using vaccines from companies he has invested in. He has funded numerous private pharmaceutical companies that manufacture vaccines and his foundation invests heavily in publicly traded pharmaceutical companies, which is a massive conflict of interest at the very least.

He has also spent fortunes on the media to cultivate a favourable image for himself and vaccines. A detailed investigation carried out by the *Colombia Journalism Review* in August 2020 showed how Bill Gates had spent at least $250 million on grants to the media.[174] This included money spent on the very 'fact-checking' organisations that defend Bill Gates and attack his detractors, the type used by Facebook to slap 'false information' strikes on patently true information.

With $250 million, one can buy a great deal of favourable media coverage, something that most people would call propaganda. Through the veneer of philanthropy, Gates has helped to shape the public discourse on public health and in particular vaccines for over a decade.

When you see a television report or a newspaper article about vaccines, the reporting organisation you are watching or reading has likely taken money from Gates' foundation. Practically any mainstream organisation you care to name has received such funding: the BBC (which laughably claims to be impartial), NBC, *Al Jazeera*, *ProPublica*, *National Journal*, *The Guardian*, *Univision*, *Medium*, *Financial Times*, *The Atlantic*, *Gannett News*, *Le Monde* and many more.

In a scarcely believable revelation, the *Colombia Journalism Review* reported that Gates had even commissioned a report in 2016 from the *American Press Institute*. The guidelines from this report were used to teach newsrooms how to demonstrate 'editorial independence from philanthropic funders'.

One can only laugh at the chutzpah.

The figure of $250 million is probably a vast underestimation as it only covers grants and not work carried out under contract. The amount of money the Bill & Melinda Gates Foundation has paid for contract work is simply not known to the public.

The bottom line is that when it comes to mainstream media reporting on Bill Gates and his projects, you will hardly ever hear criticism of Bill Gates. To criticise Gates would be biting the hand that feeds them. Another thing to remember is that many of Bill Gates' pharmaceutical partners also advertise with mainstream media. One more reason for the media not to upset Bill Gates.

If Robert F Kennedy Jr is just some crazy anti-vaxxer ranting his bizarre conspiracy theories it should be a simple matter for a multi-billionaire to take out a libel action against him and stop the accusations.

Bill Gates and his colleagues have left a trail of dead bodies in their wake. Mostly black and brown bodies and nearly always the poorest members of society. As well as the dead there are the unborn, the paralysed, the injured and the infertile. This is a simple statement of fact. Whether one believes Gates' philanthropy has done more good than harm is up to the reader to decide, but here are a couple of points to consider.

It would probably inspire greater public confidence if the WHO, UNICEF and Bill Gates prioritised clean water and sanitation above their fanatical belief that health should come from needles. Preferably we would have people pulling the levers of the international health care system who are not lifetime advocates of population reduction.

With this in mind, let us propose a formula like Bill Gates did in his 2010 TED Talk.

The more needles injected into arms = the more sterility, paralysis, misery and death.

$$NIA = ST + PAR + MIS + DE$$

If you wanted to get one of these terms down to zero – any one – what would you do?

Does any of this inspire you to take a Bill Gates' promoted vaccine? Not even one concocted with assistance from those fine scientists in the bowels of the Pentagon?

Some governments have already warned that pregnant women should not be taking COVID-19 vaccines. This sounds like sensible advice given how many miscarriages have already occurred post vaccination.[175]

None of the COVID-19 vaccines have been tested for fertility issues, so if the vaccine happened to impair or destroy one's fertility, that would simply be tough luck. No liability for the vaccine companies, no compensation and no more children for the unfortunate victims.

In the light of these sterilisation horror stories, why would anyone vaccinate a teenage child for COVID-19 when the risk of dying from COVID-19 in that age group is a rounding error?

2.40

Johnson & Johnson

Like many other rapidly promoted high fliers, Henry Kissinger was fully on board with the depopulation agenda. Would he have risen to such elevated status if he hadn't been?

Kissinger was a Nelson Rockefeller protégé. Bill Gates also has close links to the Rockefeller Foundation and, like his father, was prepared to put his money where his mouth is when it comes to the question of population numbers.

Gates has had a high-flying career and leads a charmed life it would appear, 'charmed' meaning an almost absence of criticism in mainstream media resulting from his philanthropic endeavours, not to mention legal action. It helps that the pharmaceutical companies he invests in often have legal immunity and that the UN's organisations he funds are largely free from the constraints of international law.

On the other side of the Atlantic, another father-son combination have also had very successful careers. You probably know a little bit about the son: the British Prime Minister, Boris Johnson.

His father, Stanley, is not quite as well known. Stanley Johnson is an author and a former politician (a European Member of Parliament) as well as a former bureaucrat who worked for the World Bank and European Commission. He has been writing books advocating depopulation for fifty years. His 1970 book was called *Life without Birth: A Journey Through the Third World in Search of the Population Explosion*.

Here is an extract from the introduction of that book:

Quote from the Introduction:

'Action on the population front means somehow trying to slow down or even halt this rate of growth by reducing the number of babies who are born each year. The alternative to this, both pessimists and optimists are agreed, is inevitable. If the number of new births cannot be reduced, the number of deaths each year must increase until the net growth of the world's population is cut down to a more manageable number. This business of an increase in the death rate is not, of course, attractive. The recipes - which include thermonuclear war, global famine, plague and a newly discovered horror called "ecocatastrophe" are all unpalatable. At the same time, the task of achieving a reduction in the global birth rate (which I call population control for the sake of simplicity) though clearly more death is also a form of population control'.

In Stanley Johnson's opinion, the world needs fewer births or more deaths and he refers to famines, plagues and wars as 'recipes'. Stanley Johnson proceeded to write several more books, both fiction and non-fiction, including:

The Green Revolution (1972), The Population Problem (1973-4), The Politics of Environment (1973), Pollution Control Policy of the EEC (1978), The Earth Summit: The United Nations Conference on Environment and Development UNCED (1993), World Population - Turning the Tide (1994), The Environmental Policy of the European Communities (1995), The Politics of Population: Cairo (1994), World Population and the United Nations (1987), World Population Turning the Tide (1994) and UNEP The First 40 Years; A Narrative by Stanley Johnson (United Nations Environment Programme) (2012).

You get the picture. It is easy to spot the two themes of Stanley Johnson's collected works. He is a rabid depopulation advocate and environmentalist. One could sum up his life's work in one sentence: billions of people need to die to save the planet. Now, where have we heard that before?

As to his non-fiction, one book, in particular, stands out - his 1982 book called... *The Virus*. It features a deadly pandemic and a rush for a vaccine while the US has an imbecile for a President. You cannot make this up so here is a picture of the front cover for any disbelievers.

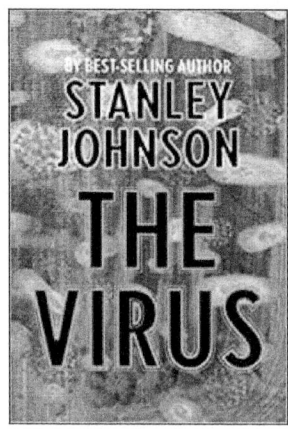

Bertrand Russell, Bill Gates, Prince Phillip, Stanley Johnson. Are these people *obsessed* with viruses?

Stanley Johnson wrote the introductory comments and edited the 1993 book *The Earth Summit*. It is a book about the Rio Earth Summit that took place in 1992.

The Rio Earth Summit of 1992 was an incredibly important event in world history. Over 100 governments attended a two-week conference to essentially pledge allegiance to the environment – a noble aim but as ever, the devil is in the detail. The summit marked the birth of *Agenda 21*, a global action plan which will end up affecting practically every living human being.

For all its enormous contents and worldwide reach, Agenda 21 has received surprisingly little mainstream scrutiny, but it probably affects your life right now in what appear to be small and innocent ways. Your local council will have its very own version of Agenda 21 whose principles derive from the original. Have you noticed cycle lanes appearing in your area and roads becoming narrower? These innocent looking changes are the work of Agenda 21. Have you noticed the endless use of the words 'sustainable' and 'equitable' in literature from your council where 'sustainable' and 'equitable' are never defined? That too, is Agenda 21 at work. The UN's 17 Sustainable Development Goals owe much to the ethos of Agenda 21. Once we see the SDGs unveiled in their true colours, it will become apparent: as far as the western world is concerned, Agenda 21 will prove to be the most detrimental policy document ever written.

That Stanley Johnson was entrusted with the task of writing the opening remarks to the official book on Agenda 21 tells us how highly he is regarded by the global policy-making classes.

2.41

Like Father, Like Son

Boris Johnson is a chip off the old block when it comes to his stated views on depopulation. In 2007, he wrote an article for *The Daily Telegraph*, also posted on his website, simply titled *Global Population Control*. In it he said:[176]

"The UN last year revised its forecasts upwards, predicting that there will be 9.2 billion people by 2050, and I simply cannot understand why no one discusses this impending calamity, and why no world statesmen have the guts to treat the issue with the seriousness it deserves."

Boris Johnson does not explain exactly why this is an 'impending calamity' (as we have seen, it isn't) but he suggests that population growth is bad for Mother Nature, bemoaning the fact that:

"We have reached the absurd position in which humanity bleats about the destruction of the environment, and yet there is not a peep in any communiqué from any summit of the EU, G8 or UN about the population growth that is causing that destruction."

Boris Johnson repeats the lie that somehow it is population growth causing the 'destruction of the environment'. This is the classic case of the fat man at the back of the boat. Worth a second look.

The growth in world population is coming from the poorest parts of the world, people who live on a few dollars a day. Poor Africans with no money are not causing the majority of the world's environmental damage or anything close to it.

It is hard to know whether these are Boris Johnson's real views because he is an inveterate liar.

Johnson was fired from *The Times* for fabricating a quote, fired as shadow Arts Minister for lying, he lied about Brexit, he lied about taxi emissions, he lied about Novichok, he lies regularly about 'Russian aggression' and he has lied from day one to the present about coronavirus. He even had the gall to tell a TV audience he never told a lie in his career.[177]

One can only imagine the lies he must tell in his personal life. Johnson has seven children that are accounted for with three different women, but chooses not to disclose the actual number of his children, meaning there are probably others. It is funny how depopulation advocates always seem to have so many children.

He is utterly shameless yet despite it all, Stanley Johnson's bumbling oaf of a son somehow found himself elevated to the position of British Prime Minister. The media could go to town with the ammunition they have on this man yet he is treated with kid gloves and portrayed as a 'character'. The media insists what a fierce intellect he is but, to anyone with eyes and ears, it is obvious he makes gaffe after gaffe which a moderately intelligent person could easily avoid. Racist gaffes too, which one would imagine could draw a huge amount of criticism from the media if the fancy ever took them.

He is not the brightest spark but certainly possesses cunning and a willingness to say the things which will further his agenda. Yet again, we see a depopulation advocate leading a charmed life when it comes to career progression, guided into high office almost as if by an invisible hand.

Earlier, we wondered if Malthus might have known his theories were lies which he promoted cynically for career reasons. It is perfectly plausible that the ambitious Boris Johnson makes these remarks with career progression in mind.

That said, when one looks back at his government's decisions to place infected COVID-19 patients into British care homes, to deny elderly people hospital treatment and the widespread 'Do Not Resuscitate' orders put in place, it makes perfect sense that a depopulation advocate was in charge.

2.42

The Club of Rome

Do you remember the name for the newly discovered horror that Stanley Johnson was so concerned about in that quotation above – 'Ecocatastrophe'?

That would certainly have been a novel word in 1970. It's almost as if somebody had just made it up. Well, that is close to exactly what had just happened. It was around this time that an entire movement was being dreamed up, one which depopulation advocates could hide behind. Paul Ehrlich also used the word 'ecocatastrophe'.

Just as eugenics proponents turned their attention to human genetic research, depopulation advocates turned to environmentalism as a means to disguise their true intentions.

In 1968, the Club of Rome was formed at the Rockefeller estate in Bellagio. The Club of Rome is an elitist think tank composed of around a hundred businessmen, scientists, former heads of state, political advisors and UN administrators. It is not related to the Catholic Church in any way, it just happened to be created in Italy.

The Club of Rome commissioned a study to be carried out by professors at MIT, the results of which led to the publication of *The Limits to Growth* in 1972. The report made predictions of imminent doom for humanity such as hunger, resource depletion and overpopulation yet they were based on computer models with assumptions that could not be verified and have since proven unrealistic, in particular the date at which the world would run out of petroleum. It looked an awful lot like Hubbert's peak but with more technical language and wild extrapolations from computer modelling.

Then, in 1974, another alarmist book titled *Mankind at the Turning Point: The Second Report to The Club of Rome* was published. The need to create a truly interdependent 'organic' society was presented as the only way to save the world. As usual with wealthy globalists, this would require countries to cede sovereignty to some supranational body or other. How about this quotation on resource allocation?

"Now is the time to draw up a master plan for organic sustainable growth and world development based on global allocation of all finite resources and a new global economic system."[178]

'Global allocation of *all* finite resources'? This 'new global economic system' sounds an awful lot like a Great Reset. 'Sustainable' too. Who would be doing the global allocating if countries are ceding their autonomy? Computer modellers? The WEF? Obviously, it would be technocrats of one flavour or another.

Overpopulation was a recurring theme in Club of Rome publications. The book's most quoted sentence is no doubt the following one:[178]

"The Earth has cancer and the cancer is man."

Blaming mankind for all of the world's ecological problems is bad enough, but to use it as a non sequitur to take global control of the world's resources while tacitly justifying depopulation is something else. (Perhaps this quote sums up the *real* turning point mankind has reached.)

The co-founder of the Club of Rome, Dr Alexander King, admitted the deception in his 1993 book *The First Global Revolution* saying:[179]

"In searching for a new enemy to unite us, we came up with the idea that pollution, the threat of global warming, water shortages, famine and the like would fit the bill... All these dangers are caused by human intervention... The real enemy, then, is humanity itself."

Please read this quote as many times as it takes to realise that we have all been conned.

This is the same rubbish that Boris Johnson, Prince Phillip and David Attenborough repeat endlessly. When you hear your friends or neighbours saying in resigned tones that the Earth is overpopulated and we are causing climate change, they are repeating this lie. They have been duped. Understand where this lie originated: the Club of Rome, an elitist think tank.

Bear in mind that the people saying this are the same people who would go on to organise the Rio Earth Summit in 1992 and to establish the UN bodies with a remit to obsess about manmade climate change such as the IPCC. We have been conditioned into hating ourselves for problems that were not our faults and in some cases, they are not even problems at all.

Laughably, the Club of Rome claims to be apolitical. Reading the foreword to Alexander King's book *The First Global Revolution* he has the absolute gall to claim:[179]

"The Club has absolutely no political ambition."

'Absolutely no political ambition', apart from agitating for a new global economic system where resources are allocated by some unnamed group of technocrats, presumably comprised of their friends.

A most insightful passage on page 73 describes the Club of Rome's thinking about globalism.

"Global governance in our vocabulary does not imply a global "government", but rather the institutions set up for co-operation, co-ordination and common action between durable sovereign states."[179]

This is from the horse's mouth. This is worth remembering when people speak of a world government or a 'one-world government'. The one-world government is something of a straw man and people dismiss the idea as fantasy. In all truth, one is unlikely ever to be formed but this would not even be necessary.

What the Club of Rome is referring to when they speak of 'global governance' is teamwork between the institutions such as the UN, the IMF the World Bank, the Bank for International Settlements (BIS) the G20, the World Trade Organisation (WTO) and the Organisation for Economic Co-operation and Development (OECD). All of these have a globalist philosophy, meaning they want to reduce and even erase national sovereignty and in some cases national borders themselves.

These institutions are supported by think tanks such as the Council on Foreign Relations (CFR), the Trilateral Commission, the Bilderberg Group and the Group of 30. In turn, they are supported by countless NGOs, including of course the WEF, many of which are funded by the big foundations and 'philanthropists'. These bodies invite the 'right' kind of people from the media, academia and big business, nearly all of them mentored into the concept of Sustainable Development.

When you look at the membership lists of these institutions, you see the same names cropping up again and again and you see the same funding sources and founding members (*cough cough* Rockefeller).

Recall how the so-called climate emergency was publicised in 2019, with Greta Thunberg and Extinction Rebellion going global almost overnight. Think back to 2020, when multiple western governments simultaneously decided to implement lockdowns based on the models of a discredited computer modeller based at Imperial College in the UK. What are the odds that all these national governments independently decided to eschew their national brainpower and rely on a man called Neil Ferguson with possibly the world's worst track record in forecasting? Yes, the odds are zero. It was coordinated.

A world government was not required to achieve this. As to 'global governance', well in this author's opinion, it is already here. The 'co-operation, coordination and common action between durable sovereign states' the Club of Rome hankered for is obvious to see, but I wouldn't call it progress. Nor would I class the western states as particularly 'sovereign'. As to 'durable', well let's wait and see. When one assesses the US either politically, economically or socially, all the signs are telling us it is circling the drain.

On page 71, the authors lament: "In its present form, democracy is no longer well suited for the tasks ahead."[179]

That was certainly true for the 'tasks ahead' that the Club of Rome had in mind. Probably just a disinterested observation though, seeing the Club of Rome has 'absolutely no political ambition'.

This demonstrates how much trust the Club of Rome deserves. One should assume that anything emerging from this group is designed to deceive and for their faithful followers to repeat. It is also interesting to note how often the Club of Rome's announcements become policy or are adopted as law in various countries and repeated by other think tanks and NGOs.

To wit, on 24 November 2018, the Club of Rome issued a call for immediate emergency action with 'We Don't Have Time', the social network established by businessman Ingmar Rentzhog as part of a slick climate change PR campaign.[180]

"The Club of Rome and its partners call on all stakeholders – governments, civil society, scientific institutions, and business – to adopt the following emergency action plan."[181]

No political ambition there then.

2.43

Environmentalism All the Way

The Club of Rome's message has been copied and pasted all over the world. Governmental organisations, the UN and the WEF embody the spirit of the Club of Rome. The Green New Deal and Sustainable Development Goals are based on its founding principle and environmental groups like XR preach the gospel of climate change as a clear and present danger. Just as the Club of Rome intended, humanity is portrayed as the enemy.

Carbon dioxide is demonised as a pollutant although it is nothing of the sort. The attacks on CO_2 have morphed into attacks on carbon itself. Human beings are a carbon-based life form so perhaps this is the whole point. If humanity is bad it stands to reason that depopulation must be good, or even necessary.

Here is national treasure Sir David Attenborough again, spelling it out:

"All our environmental problems become easier to solve with fewer people, and harder – and ultimately impossible – to solve with ever more people."

The message could not be any clearer. The population reduction agenda is not a hidden one. It is out in the open. Attenborough's quote appears on the front page of the website 'populationmatters.org', a UK-based charity that he is a patron of.[182] Population bomber Paul Ehrlich is also a patron and Prince Charles was formerly one too. The only purpose of this British charity is to promote a lower world population. On the website, a large counter is displayed with a running total of the world population creeping upwards. Alarmist articles stress the need for population growth to fall to save the environment.

In 2009, a project called 'PopOffsets' was launched where people were invited to offset their CO_2 emissions by paying for family planning programmes in Africa. People could pay to alleviate their environmental guilt by preventing Africans from being born!

It appears people are walking among us who believe they can make good on their 'sinful' CO_2 output by funding sterilisations so that several African children never get to be born. Imagine thinking these sick thoughts. Imagine encouraging this kind of thinking.

I'm afraid this is what Sir David Attenborough really stands for.

The name of this programme was changed from PopOffsets to 'Empower to Plan' in 2018.[183] Globalists and their minions love using the word 'empower' in their propaganda almost as much as they love saying 'diversity' (but not as much as 'sustainable'). However, this name change might have come about because even these people realised how desperately offensive the notion of offsetting your consumption with actual human lives sounded.

If the people who signed up for this scheme were not so brainwashed with green propaganda, they would realise there was no need to feel such environmental guilt to begin with. All they are succeeding in doing is preventing a bunch of kids from being born, something they probably feel smug about in the misguided belief they are helping the planet.

The website might as well be called eugenicsmatters.org. And this organisation is a charity.

2.44

Green Foreign 'Aid'

How many British taxpayers are aware that the UK contributed tens of millions of pounds of taxpayer funds to India's sterilisation programme in the name of climate change? The programme had horrendous consequences. *The Guardian* reported in 2012 that:[184]

"... a working paper published by the UK's Department for International Development in 2010, cited the need to fight climate change as one of the key reasons for pressing ahead with such programmes. The document argued that reducing population numbers would cut greenhouse gases, although it warned that there were 'complex human rights and ethical issues' involved in forced population control."

The idea that sterilising poor Indian women will 'fight climate change' is such a stretch of the imagination one wonders if even the report's authors believed what they were writing. NGO workers were given cash bonuses to perform sterilisations and Indian women bribed and cajoled to undergo operations with as little as 600 rupees and a sari. As *The Guardian* reported:

"... in Bihar private clinics receive 1,500 rupees for every sterilisation, with a bonus of 500 rupees a patient if they carry out more than 30 operations on a particular day. NGO workers who convince people to have the operations receive 150 rupees a person, while doctors get 75 rupees for each patient."[184]

Conditions were ripe for abuse and of course, that is what happened.

One DfID-funded doctor performed 53 sterilisations in just two hours by flashlight and botched all 53. He did not even sterilise his instruments between operations because he was in such a rush to make as much money as he could. An Indian court heard how a 35-

year-old woman who was pregnant with twins at the time bled to death. Are we really to believe these policies are paid for by taxpayers thousands of miles away to 'cut greenhouse emissions'?

Next time you hear someone ranting about the scandalous sums Britain spends on foreign aid perhaps point out to them the recipients of the 'aid' are not exactly benefitting from their taxes.

As to DfID's comment that the policy raised 'complex human rights and ethical issues', do these issues seem particularly complex to you on an ethical level, or is the policy plainly wrong?

DfID said the sterilisation programme was to reduce greenhouse gas emissions and then provided the money and incentives for these poor women to be injured in filthy conditions as they were made sterile. The only benefit DfID could point to is fewer molecules of CO_2 entering the atmosphere which wouldn't make the slightest difference to global warming in any case. Perhaps we could just drop fewer bombs in the Middle East instead?

The only conclusion one can draw is that after hemming and hawing about 'ethical considerations', DfID believes these Indian lives are a price worth paying for flimsy environmental reasons.

This is the problem with junk science.

Voltaire was probably talking about religion when he wrote 'whoever can make you believe absurdities can make you commit atrocities' but he might as well have been talking about climate change zealotry in the modern era.

2.45

Mad Science

Speaking of absurdities in the name of environmentalism, scientists are considering attempts to cool the planet by spraying reflective dust in the skies. In 2009, Obama's Malthusian science czar, John Holdren, said that the planet could be cooled using geoengineering and the US Congress held hearings on the matter shortly afterwards.[185]

The idea would be to fill the sky with particles that reflect heat into space and reduce the temperature on earth. Bill Gates has echoed Holdren's ideas more recently and has funded a Harvard University research project called SCoPEx (Stratospheric Controlled Perturbation Experiment) which has the same aims.[186]

The project is currently on hold but there are good reasons to believe we haven't heard the last of this lunacy. Even if it did work as advertised, it would be to the detriment of millions. Cooling the planet would cause harm to harvests and raise heating costs. As to the unintended consequences, God knows. It's not as if these megalomaniacs even asked before they conduct their mad science experiments that affect the entire world. Just as nobody asked for public consent for Elon Musk's Starlink which will bathe the globe in electromagnetic radiation from satellites, neither will the public have a vote on geoengineering. Do not be surprised to see this insane plan introduced at some point, even during a solar minimum.

The US Office of the Director of National Intelligence hinted as much in its *Global Trends* report published in March 2021. This is a report issued every four years that looks 20 years into the future from a US perspective and examines different global scenarios. It is full of the usual alarmism that climate change is going to cause food shortages which will lead to uncertainty, strife and... increased nationalism!

Seemingly, there isn't a problem facing humanity that does not lead to the non-sequiturs of increased nationalism, white supremacy or domestic terrorism in the eyes of those who write policy documents for US government agencies. However, on page 38 there is an interesting quote about geoengineering in their scenario planning.[187]

"... as the world gets closer to exceeding 1.5°C – probably within the next 20 years – calls will increase for geoengineering research and possible deployment to cool the planet, despite possibly dire consequences."

It is interesting that the authors note the 'possibly dire consequences' but do not outright condemn the schemes, even though the risks of interfering with a natural system are unknown. In one of their 'this-is-how-it-could-go-wonky' disaster scenarios for 2040, it states:

"Lacking coordinated, multilateral efforts to mitigate emissions and address climate changes, little was done to slow greenhouse gas emissions, and some states experimented with geoengineering with disastrous consequences."

You can see where this narrative is going: because selfish humans did not cut back on their sinful carbon dioxide emissions, some people just had to commence geoengineering, as you do. But then some bad stuff happened and people got angry, so either we all need to stop driving and eating meat etc or else someone will have to start spraying stuff in the skies to save us. However, if geoengineering is to be done it needs to be agreed at a global level, not done by some random state actor. What is required is a fine group of unelected technocrats to assume global control. We need more globalism to save us from ourselves!

The bulk of the 150-page report is predicated on the dangers of climate change, specifically with the climate getting *warmer*, something they assert is being caused by man (humanity being the enemy and all that). All this is as the world is entering a solar minimum. The absolute insanity.

Perhaps it is not insanity and the report has a different objective: to seed 'solutions'. A feature of the solar minimum is that as we enter it, there is likely to be a flurry of extreme weather events. There could be cold snaps, hot spells, heavy rainfall, tornados, giant hail storms and so on. Guess what all this gets blamed on by people brought up on climate change propaganda?

The answer is simple when you've grown up to believe that the enemy is man.

Even when temperature is falling, the climate change disciples will forget about 'global warming' and just call it climate change. It is not impossible to envisage a future scenario where a brainwashed public even ends up supporting a war with a country they were told had committed crimes against the environment. *Foreign Policy* magazine – the trusty mouthpiece of the immensely influential Council on Foreign Relations – already went up with a trial balloon in 2019 when the Amazon was burning, asking, "Do states have the right – or even the obligation – to intervene in a foreign country in order to prevent it from causing irreversible and possibly catastrophic harm to the environment?"[188]

Sometimes these policy documents are telling us things. There was even a link to this obscure DNI report in *The Daily Mail* online, almost as if someone wanted the public to see it. Do you think perhaps the US intelligence agency knows something we don't about the future?

I'll tell you what I think is happening here, just a personal opinion. I believe the authors of this report understand fine well that climate change is not caused solely by man. I also believe that these sky spraying activities will not lower the global temperature by any meaningful amount, if at all. But the solar minimum *will* lower the temperature, something I believe the authors are also aware of. (And they ought to be aware, working for the Director of National Intelligence.) At some point, sky spraying activities will commence publicly and it will be claimed they are working, meaning they are lowering the temperature to save us from ourselves. As this is happening, shorter growing seasons will mean food shortages and fuel poverty, but we will be led to believe the increased hunger, fuel poverty and death is for 'the greater good' and besides, most people don't care about other people dying these days. What they will be spraying in the sky is anyone's guess but it won't be the thing that is lowering the temperature. Whatever political capital the powers-that-be choose to make of it will depend on how the ruling class want to drive global change, and the environmentalists will believe every last word that comes out of their television sets, dictated to them by newsreaders quoting from military intelligence sources. The desired global change will probably be... well you know what it is that drives these people by now.

2.46

Final Thoughts

The world is not overpopulated. This is the biggest lie of all and it is perpetuated to persuade us to accept those other heinous lies: that mankind is causing climate change and that mankind is the enemy.

It isn't true. We have enough space, we have enough food and we have enough resources for the time being, not to mention unquantifiable human ingenuity. Nor will manmade climate change kill us all any time soon. Climate change *could* kill us all one day but it would be 'death by natural causes' were that to happen. Nothing we can do about it. This is good and bad news.

People who have been walking around with the weight of the world on their shoulders, wrestling with the dissonance of wanting to be a good person while also wanting billions to die, can throw off these shackles.

Seeing through the lie comes at a price though, and this is the bad news. One comes to understand why it is we have been lied to. To be blunt, it is because the oligarchy does not want us inhabiting this fine planet.

Does that mean they plan on doing anything about it? Should we be worried? That is something we can only guess because nobody is sensibly going to admit to a plan to kill large numbers of people. Well, apart from really high-minded types perhaps, the sort who get to work with Presidents and win awards and receive copious grant money, but even then not in public. Look around the world today and what do you see?

- Mass vaccination where there is no need to vaccinate, with experimental gene therapy at that.

- Economic 'assistance' policies that will cause economic harm and centralise power even further.

- Unnecessary lockdowns that damage growth and disrupt supply chains, causing hunger and death as preventable diseases go untreated.
- Geoengineering plans when there is no need to reduce the sun's output reaching Earth as we enter a solar minimum.
- Feverish interest in human genetics yet a complete lack of enthusiasm in actually promoting human health naturally.
- People making alarmist predictions being promoted no matter how badly they get it wrong, particularly with computer modelling (climate change, COVID-19).
- An 'anything goes' attitude to climate solutions, whatever the cost and whatever the consequences to innocent people, usually innocent, poor people.

Where we witness these things we seem to find the same people: billionaires, philanthropists, and generally elitist types who believe the earth is overpopulated. Think Club of Rome. Think eugenics. Did it ever go away?

It is unlikely that our elected politicians care whether we live or die. If you have finished your working career and don't pay any taxes it is a safe bet they *want* you to die. It should be obvious by now that the UN cares even less about your life, being unelected and far away, and with a predilection for reducing human fertility.

We should get used to the endless propaganda demeaning human life because it is only going to get worse in the coming decades. Expect to see more promotion of abortion and euthanasia, more calls to end life at 75[189], with feel-good stories about the benefits of composting human bodies[190], the widespread use of 'Do Not Resuscitate' orders, more crazy talk of 'eating humans to save the planet'[191], as well as 'not having children at all to save the planet', and pretty much anything else that devalues human life.

When you see these stories, see them for what they are.

ENDNOTES:

1 - https://www.politics.co.uk/news/2013/01/22/david-attenborough-humans-are-a-plague-on-the-earth/
2 - http://web.archive.org/web/20210205161840/https://www.worldometers.info/world-population/
3 - http://www.newgeography.com/content/001689-how-much-world-covered-cities
4 - http://metrocosm.com/world-population-split-in-half-map/
5 - https://www.un.org/development/desa/en/news/population/2018-revision-of-world-urbanization-prospects.html
6 - https://www.researchgate.net/publication/241746569_We_Already_Grow_Enough_Food_for_10_Billion_People_and_Still_Can't_End_Hunger
7 - https://borgenproject.org/the-cost-to-end-world-hunger/
8 - https://eurymanthus.wordpress.com/2021/05/23/paying-farmers-not-to-grow-crops-not-just-the-us-but-what-about-food-price-inflation-shortages/
9 - https://www.theodysseyonline.com/13-reasons-why-we-need-california-agriculture
10 - http://www.williamengdahl.com/englishNEO10June2021.php
11 - https://healthimpactnews.com/2021/california-state-water-board-is-manufacturing-a-drought-by-draining-reservoirs-into-the-ocean/
12 - http://web.archive.org/web/20141027200630/https://populationmatters.org/documents/rsa_attenborough.pdf
13 - https://www.worldbank.org/en/topic/poverty/overview
14 - https://www.theguardian.com/business/2004/jul/29/oilandpetrol.news
15 - https://www.forbes.com/sites/rrapier/2019/02/14/how-much-oil-does-saudi-arabia-have/
16 - https://www.reuters.com/article/uk-saudi-economy-reserves-analysis-idUKKBN19I17R
17 - https://www.investopedia.com/terms/h/hubbert-peak-theory.asp
18 - https://www.bp.com/content/dam/bp/business-sites/en/global/corporate/pdfs/energy-economics/statistical-review/bp-stats-review-2020-full-report.pdf
19 - https://www.nytimes.com/2021/02/12/business/dealbook/shell-peak-oil.html
20 - https://edition.cnn.com/2021/02/11/business/shell-oil-production-peak/index.html
21 - https://www.thegwpf.com/methane-hydrates-fossil-fuel-or-energy-saviour/
22 - https://www.investopedia.com/terms/h/hubbert-peak-theory.asp
23 - https://www.world-nuclear.org/information-library/current-and-future-generation/plans-for-new-reactors-worldwide.aspx
24 - https://www.world-nuclear.org/information-library/country-profiles/countries-o-s/russia-nuclear-power.aspx
25 - https://cluborlov.blogspot.com/2019/01/the-future-of-energy-is-bright-part-ii.html
26 - https://www.sciencealert.com/for-the-first-time-ever-scientists-create-diamonds-in-the-lab-without-heat
27 - http://mattslay.com/what-percentage-of-the-suns-energy-hits-the-earth/
28 - http://www.geocities.ws/popcom10/calcu.htm

29 - http://web.archive.org/web/20200701085904/https://www.cia.gov/library/publications/the-world-factbook/fields/351.html
30 - https://www.worldometers.info/world-population/#growthrate
31 - http://web.archive.org/web/20210223123202/https://www.worldometers.info/world-population/population-by-country/
32 - https://www.thelancet.com/pdfs/journals/lancet/PIIS0140-6736(20)30977-6.pdf page 1178
33 - http://www.newgeography.com/content/005525-death-spiral-demographics-the-countries-shrinking-the-fastest
34 - https://population.un.org/wpp/Publications/Files/WPP2019_Methodology.pdf
35 - https://www.nasa.gov/press-release/solar-cycle-25-is-here-nasa-noaa-scientists-explain-what-that-means/
36- https://www.bibliotecapleyades.net/sociopolitica/esp_sociopol_depopu109.htm
37 - https://www.ted.com/talks/bill_gates_innovating_to_zero/transcript
38 - https://www.bibliotecapleyades.net/sociopolitica/esp_sociopol_depopu109.htm
39 - https://www.independent.co.uk/news/royals-shooting-passion-draws-bad-blood-1315283.html
40 - http://twainquotes.com/
41 - https://research.calvin.edu/german-propaganda-archive/goeb29.htm
42 - https://www.youtube.com/watch?v=3rWU_VDa1Js&ab_channel=ScootleRoyale
43 - https://www.youtube.com/watch?v=taoHk_enqWA
44 - https://www.express.co.uk/news/royal/1095825/prince-philip-overcrowding-prince-william-prince-charles-kate-middleton-royal-news-spt
45 - Reported by Deutsche Press Agentur (DPA), August, 1988
46 - Preface to Down to Earth by HRH Prince Philip, Duke of Edinburgh, 1988, p8
47 - https://www.econlib.org/library/Malthus/malPlong.html?chapter_num=47#book-reader
48 - http://www.econlib.org/library/Malthus/malPlong30.html#firstpage-bar
49 - https://www.econlib.org/library/Malthus/malPlong.html?chapter_num=47#book-reader
50 - The life, letters and labours of Francis Galton - https://archive.org/details/b29000695_0003/page/n11/mode/2up
51 - https://www.galtoninstitute.org.uk/sir-francis-galton/eugenics-and-final-years/
52 - https://www.telegraph.co.uk/news/2020/11/10/spreading-anti-vaxx-myths-should-made-criminal-offence/
53 - https://royalsociety.org/news/2020/11/vaccine-hesitancy-threatens-to-undermine-pandemic-response/
54 - https://royalsociety.org/-/media/policy/projects/set-c/set-c-vaccine-passports.pdf
55 - https://theinternationalforecaster.com/topic/international_forecaster_weekly/the_ugly_truth_about_the_minimum_wage
56 - https://pubs.aeaweb.org/doi/pdf/10.1257/089533005775196642
57 - Webb, Sidney.1912. "The Economic Theory of a Legal Minimum Wage."Journal of Political Economy. December, 20:10, pp. 973–98
58 - Meeker, Royal.1910. "Review of Cours d'Economie Politique."Political Science Quarterly.25:3, pp. 543–45

59 - Seager, Henry Rogers.1913a. "The Minimum Wage as Part of a Program for Social Reform."Annals of the American Academy of Political and Social Science. July, 48, pp. 3–12

60 - https://www.macrotrends.net/countries/USA/united-states/life-expectancy

61 - https://www.commondreams.org/news/2010/03/24/weight-poor-strategy-end-poverty

62 - https://www.ukpressonline.co.uk/ukpressonline/open/simpleSearch.jsp;jsessionid=955FA34B50F13C71BA66870AC12D5C12?is=1

63 - https://www.youtube.com/watch?v=Ymi3umIo-sM

64 - https://www.nobelprize.org/prizes/literature/1925/shaw/facts/

65 - https://en.wikipedia.org/wiki/International_Eugenics_Conference#cite_note-Pearson-1

66 - https://www.ippf.org/Family-Planning-London

67 - https://www.americanthinker.com/articles/2012/06/melinda_gates_talks_eugenics.html

68 - https://www.hli.org/2012/10/opening-the-gates-wide-to-population-control-abuse/

69 - https://www.who.int/pmnch/media/news/2012/20120627_family_planning_summit/en/index2.html

70 - United Nations Population Fund (UNFPA). Financial Resource Flows for Population Activities [annual reports]. Table A.1, "Primary Funds of Donor Countries for Population Assistance, by Channel of Distribution." For complete details and calculations, see Excel spreadsheet F-18-05.XLS.

71 - https://www.who.int/pmnch/about/steering_committee/b12-12-item5_fp_summit.pdf

72 - Edwin Black – War against The Weak: Eugenics and America's Campaign To Create a Master Race (2% kindle edition)

73 - Edwin Black – War against The Weak: Eugenics and America's Campaign To Create a Master Race (49% kindle edition)

74 http://hnn.us/articles/1796.html

75 - Edwin Black – War against The Weak: Eugenics and America's Campaign To Create a Master Race (40% kindle edition)

76 - https://openlibrary.org/books/OL13516132M/The_passing_of_the_great_race

77 - http://hnn.us/articles/1796.html

78 - https://ia800709.us.archive.org/7/items/b30619385/b30619385.pdf

79 - https://www.law.cornell.edu/supremecourt/text/274/200

80 - Edwin Black – War against The Weak: Eugenics and America's Campaign To Create a Master Race (2% kindle edition)

81 - Ruth Engs - The eugenics movement: An encyclopedia p102.

82 - http://hnn.us/articles/1796.html

83 - Edwin Black – War against The Weak: Eugenics and America's Campaign To Create a Master Race (44% kindle edition)

84 - Edwin Black – War against The Weak: Eugenics and America's Campaign To Create a Master Race (54% kindle edition)

"Wehrmachtsauftragsnummer: S 4891-5378." Posner and Ware, p. 18.

85 - Edwin Black – War against The Weak: Eugenics and America's Campaign To Create a Master Race (53% kindle edition)

86 - Edwin Black – War against The Weak: Eugenics and America's Campaign To Create a Master Race (57% kindle edition)

87 - Michael C Carroll - Lab 257: The Disturbing Story of the Government's Secret Germ Laboratory
88 - L. Fletcher Prouty "The CIA, Vietnam and the Plot to Assassinate John F. Kennedy"
89 - Wall Street and the Rise of Hitler: The Astonishing True Story of the American Financiers Who Bankrolled the Nazis - Anthony C Sutton
90 - http://www.theinsider.org/news/article.asp?id=0605
91 - http://www.digitaljournal.com/news/world/bush-grandpa-traded-with-enemy-for-3-years-before-assets-seized/article/424715
92 - https://www.nyu.edu/projects/sanger/articles/bush_family_planning.php
93 - Nazi International - Joseph P Farrell (p10)
94 - Edwin Black - IBM and the Holocaust: The Strategic Alliance between Nazi Germany and America's Most Powerful Corporation
95 - https://en.wikipedia.org/wiki/History_of_IBM
96 - https://uk.news.yahoo.com/moderna-collaborate-ibm-covid-19-072552443.html
97 - https://www.modernatx.com/ecosystem/strategic-collaborators/foundations-advancing-mrna-science-and-research
98 - https://investors.modernatx.com/news-releases/news-release-details/darpa-awards-moderna-therapeutics-grant-25-million-develop
99 - https://singularityhub.com/2018/01/25/heres-the-tech-that-could-one-day-track-boost-or-erase-human-memory/amp/
100 - https://www.theguardian.com/science/2017/dec/04/us-military-agency-invests-100m-in-genetic-extinction-technologies
101 - https://www.gatesnotes.com/Health/What-you-need-to-know-about-the-COVID-19-vaccine
102 - https://www.ema.europa.eu/en/documents/product-information/covid-19-vaccine-astrazeneca-product-information-approved-chmp-29-january-2021-pending-endorsement_en.pdf
103 - https://www.hek293.com/
104 - https://blogs.bmj.com/bmj/2021/01/04/peter-doshi-pfizer-and-modernas-95-effective-vaccines-we-need-more-details-and-the-raw-data/
105 - https://financialpost.com/financial-times/why-the-pfizer-ceo-selling-62-of-his-stock-the-same-day-as-the-vaccine-announcement-looks-bad
106 - https://www.justice.gov/opa/pr/justice-department-announces-largest-health-care-fraud-settlement-its-history
107 - https://www.dmlawfirm.com/crimes-of-covid-vaccine-maker-pfizer-well-documented/
108 - http://labeling.pfizer.com/ShowLabeling.aspx?id=14471
109 - https://www.clinicaltrials.gov/ct2/show/NCT04470427
110 - https://www.clinicaltrials.gov/ct2/show/NCT04368728
111 - https://www.clinicaltrials.gov/ct2/show/NCT04516746
112 - https://childrenshealthdefense.org/defender/tucker-carlson-how-many-americans-died-covid-vaccines/
112i - https://childrenshealthdefense.org/defender/mainstream-media-fda-approval-pfizer-vaccine/
112ii - https://www.law.cornell.edu/uscode/text/21/360bbb-3
113 - https://www.forbes.com/sites/mattperez/2020/03/18/bill-gates-calls-for-national-tracking-system-for-coronavirus-during-reddit-ama/#7ba3e6c36a72

114 - https://stm.sciencemag.org/content/scitransmed/11/523/eaay7162.full.pdf
115 - https://unlimitedhangout.com/2020/12/investigative-series/developers-of-oxford-astrazeneca-vaccine-tied-to-uk-eugenics-movement/
116 - https://www.biogeneus.com/blog/nanobots-future-trend-in-drug-delivery-and-therapeutics/
117 - https://www.nist.gov/system/files/documents/pml/div683/bioelectronics_report.pdf
118 - https://www.weforum.org/agenda/2020/06/internet-of-bodies-covid19-recovery-governance-health-data/
119 - http://web.archive.org/web/20200925125930/https://bwn.ece.gatech.edu/papers/2015/j3.pdf
120 - https://www.thelancet.com/pdfs/journals/lanplh/PIIS2542-5196(18)30221-3.pdf
121 - http://web.archive.org/web/20040718231114/http://www.arlingtoninstitute.org/library/Small%20Security.pdf
122 - https://www.gatesfoundation.org/How-We-Work/Quick-Links/Grants-Database/Grants/2014/01/OPP1068198
123 - https://nationalpost.com/news/bill-gates-funds-birth-control-microchip-that-lasts-16-years-inside-the-body-and-can-be-turned-on-or-off-with-remote-control
124 - https://www.youtube.com/watch?v=CLUWDLKAF1M&ab_channel=CNET
125 - https://www.independent.co.uk/life-style/gadgets-and-tech/elon-musk-monkey-neuralink-video-b1828913.html
126 - https://summit.news/2020/11/16/klaus-schwab-great-reset-will-lead-to-a-fusion-of-our-physical-digital-and-biological-identity/
127 - Klaus Schwab; Satya Nadella; Nicholas Davis. Shaping the Future of the Fourth Industrial Revolution (Kindle Location 9%). Kindle Edition.
128 - Klaus Schwab; Satya Nadella; Nicholas Davis. Shaping the Future of the Fourth Industrial Revolution (Kindle Location 6%). Kindle Edition.
129 - Klaus Schwab; Satya Nadella; Nicholas Davis. Shaping the Future of the Fourth Industrial Revolution (Kindle Location 9%). The Crown Publishing Group. Kindle Edition.
130 - https://www.modernatx.com/mrna-technology/mrna-platform-enabling-drug-discovery-development
131 - Consider the Lilies: A Review of 18 Cures for Cancer and Their Legal Status – Mary Maxwell
132 - https://web.archive.org/web/20131221082142/http://www.jeffreyepsteinforum.com/2012_08_01_archive.html
133 - http://web.archive.org/web/20190812055633/https://www.nytimes.com/2019/07/31/business/jeffrey-epstein-eugenics.html
134 - https://www.theguardian.com/commentisfree/2019/sep/07/jeffrey-epstein-mit-funding-tech-intellectuals
135 - https://web.archive.org/web/20120507122312/http://www.jeffreyepsteinfoundation.com/apps/photos/photo?photoid=154263327
136 - http://web.archive.org/web/20190801000630/https://www.nytimes.com/2019/07/31/business/jeffrey-epstein-eugenics.html

137 - https://www.michaeleisen.org/blog/?p=1825
138 - https://www.buzzfeednews.com/article/peteraldhous/eric-lander-biden-science-adviser
139 - https://www.theguardian.com/us-news/2021/jan/16/joe-biden-scientific-advisers-eric-lander-cabinet
140 - Bertrand Russell, The Impact of Science on Society, 1953 p103
https://archive.org/details/TheImpactOfScienceOnSociety-B.Russell/page/n53/mode/2up
141 - The Population Bomb, Paul Ehrlich, 1968
142 - The Population Bomb, Paul Ehrlich 1970 Edition
143 - Paul Ehrlich, quoted in Julian Simon, The Ultimate Resource 2, (Princeton: Princeton University Press, 1996), p. 35.
https://www.masterresource.org/holdren-john/halloween-hangover/
144 - Ecoscience - Population, Resources, Environment, p787-788
145 - Ecoscience - Population, Resources, Environment, p786-7
146 - Ecoscience - Population, Resources, Environment, p942-94
147 – Paul Ehrlich, The Machinery of Nature, Simon & Schuster, New York, 1986, p. 274.
https://www.masterresource.org/holdren-john/john-holdren-on-global-warming-part-ii-in-a-series-on-obamas-new-science-advisor/
148 - https://en.wikipedia.org/wiki/Anthony_Fauci#cite_note-15
149 - https://pdf.usaid.gov/pdf_docs/Pcaab500.pdf
150 - https://www.democracynow.org/2014/4/4/is_usaid_the_new_cia_agency
151 - http://web.archive.org/web/20141112092221/http://www.kccb.or.ke/home/news-2/press-statement-5/
152 - https://www.lifesitenews.com/news/a-mass-sterilization-exercise-kenyan-doctors-find-anti-fertility-agent-in-u
153 - https://www.who.int/immunization/monitoring_surveillance/resources/milstien.pdf
154 - https://pubmed.ncbi.nlm.nih.gov/12346214/
155 - http://web.archive.org/web/20180204015624/http://apanews.net/en/pays/kenya/news/kenya-thousands-infertile-after-govt-sponsored-vaccination-odinga
156 - https://healthimpactnews.com/2013/no-polio-in-the-philippines-since-1993-but-mass-polio-vaccination-program-started-among-500000-typhoon-victims/
157 - https://www.unhcr.org/news/press/2013/11/52838f036/unhcr-teams-working-fight-expansion-polio-syria.html
158 https://pubmed.ncbi.nlm.nih.gov/18180708/
159 - http://jid.oxfordjournals.org/content/192/12/2033.full.pdf
160 - https://www.healio.com/news/pediatrics/20121212/idc1212syriavax_10_3928_1081_597x_20121201_12_976360
161 - https://apnews.com/article/d954ca949bf34124bc980d4e2f732b16
162 - https://www.icandecide.org/ican-vs-unicef/
163 - https://web.archive.org/web/20190507211443/https://icandecide.org/wp-content/uploads/whitepapers/Unicef-DTP.pdf
164 - https://www.ncbi.nlm.nih.gov/pmc/articles/PMC5360569/
165 - https://pubmed.ncbi.nlm.nih.gov/?term=PETER+AABY%5BAuthor+-+Full%5D
166 - https://www.forbes.com/sites/eliseknutsen/2013/01/28/israel-foribly-injected-african-immigrant-women-with-birth-control/

167 - https://www.independent.co.uk/news/world/middle-east/israel-gave-birth-control-ethiopian-jews-without-their-consent-8468800.html
168 - https://www.camera.org/article/camera-prompts-ha-aretz-correction-on-ethiopian-birth-control-story/
169 - https://www.ncregister.com/news/fujimori-re-imprisonment-and-peru-s-forgotten-forced-sterilization-program
170 - https://mises.org/wire/forced-sterilizations-peru-paid-us-taxpayers
171 - https://www.nlm.nih.gov/nativevoices/timeline/543.html
172 - https://childrenshealthdefense.org/news/government-corruption/gates-globalist-vaccine-agenda-a-win-win-for-pharma-and-mandatory-vaccination/
173 - https://abcnews.go.com/Technology/wireStory/rfk-jr-kicked-off-instagram-vaccine-misinformation-75830042
174 - https://www.cjr.org/criticism/gates-foundation-journalism-funding.php
175 - https://openvaers.com/covid-data/reproductive-health
176 - https://www.boris-johnson.com/2007/10/25/global-population-control/
177 - https://www.independent.co.uk/news/uk/politics/boris-johnson-lie-career-general-election-brexit-itv-a9225601.html
178 - Mankind at the turning point: the second report to the Club of Rome
179 - https://wakeup-world.com/wp-content/uploads/2015/12/The-Council-of-The-Club-of-Rome-The-First-Global-Revolution.pdf pages 71-75
180 - https://climateemergencydeclaration.org/club-of-rome-climate-emergency-plan/
181 - https://clubofrome.org/wp-content/uploads/2018/10/COR_Climate-Emergency-Plan-.pdf
182 - https://populationmatters.org/
183 - https://en.wikipedia.org/wiki/PopOffsets
184 - https://www.theguardian.com/world/2012/apr/15/uk-aid-forced-sterilisation-india
185 - https://www.technologyreview.com/2009/12/21/207045/the-geoengineering-gambit/
186 - https://www.forbes.com/sites/arielcohen/2021/01/11/bill-gates-backed-climate-solution-gains-traction-but-concerns-linger/
187 - https://www.dni.gov/files/ODNI/documents/assessments/GlobalTrends_2040.pdf
188 - https://foreignpolicy.com/2019/08/05/who-will-invade-brazil-to-save-the-amazon/
189 - https://www.theatlantic.com/magazine/archive/2014/10/why-i-hope-to-die-at-75/379329/
190 - https://www.sierraclub.org/sierra/now-you-can-compost-human-bodies-too
191 - https://summit.news/2019/09/04/swedish-behavioral-scientist-suggests-eating-humans-to-save-the-planet/

About the Author

Will Rowlands (pronouns: off, sod) is a heterosexual white male whose thoughts and opinions are motivated solely by far-right extremism and bigotry (well duh). He writes conspiracy nonsense because his cheque from the Kremlin hasn't cleared yet. In reality he reads the Guardian while drinking organic fair trade soy lattes and weeping about climate change. This is his second book.

Also author of "The World in Bullshit - COVID19: Setting the Trap"

Printed in Great Britain
by Amazon